Arthropod Bioacoustics:
Neurobiology and Behaviour

Arthropod Bioacoustics
Neurobiology and Behaviour

ARTHUR W. EWING

Comstock Publishing Associates, a division of
Cornell University Press | Ithaca, New York

Library of Congress Cataloging-in-Publication Data

Ewing, Arthur M.
    Arthropod bioacoustics; neurobiology and behaviour/Arthur W.
Ewing.
        p.   cm.
    Includes bibliographical references.
    ISBN 0-8014-2478-X (alk. paper)
    1. Sound production in insects.   2. Insects—Physiology.
3. Insects—Behavior.   4. Arthropods—Physiology.   5. Arthropods—
Behavior.   6. Bioacoustics.   I. Title.
    [DNLM: 1. Acoustics.   2. Arthropods—Physiology.   3. Nervous
System—physiology.   4. Vocalization, Animal.   QL 434.72 E95a]
QL496.5.E96   1989
595′.2041825—dc20
DNLM/DLC
for Library of Congress

First published 1989 by Cornell University Press and Edinburgh University Press

International Standard Book Number 0-8014-2478-X.
Library of Congress Catalog Card Number 89-81495.

Printed in Great Britain

Contents

Preface

The sounds produced by arthropods, in particular the songs of insects, are among the major sources of biological noise in the environment. In many tropical forests and in the pine-woods of the Mediterranean region, it is the songs of cicadas, crickets and bush-crickets that are dominant, and not the calls of birds or other vertebrates. This is not so true of temperate regions, but in biotypes like alpine meadows in summer one can hear the continuous stridulation of grasshoppers and crickets. It is not surprising therefore that biologists and naturalists alike for more than two centuries have shown an interest in insect songs. Over and above those insects whose songs are sufficiently loud for us to hear unaided, there are uncounted numbers of species passing quiet, intimate messages or transmitting their signals in the form of vibrations through the substrate. Of the latter, some, like those of the Death-watch Beetle, can be heard, but we remain unaware of the great majority. The acoustic barrier between water and air cuts us off from other sources of biological signals. In shallow tropical seas, the loudest sounds are often the clicks produced by large numbers of Pistol Shrimps, and in freshwater streams and lakes there are beetles, water bugs, caddis and dragonfly larvae which communicate via underwater sound.

Technical developments over recent years have opened up these hitherto hidden worlds, and it is now possible to listen in to and record these signals relatively simply. It is noticeable that entomological books aimed primarily at the amateur often contain detailed descriptions of insect song and use these for identification. It is even possible to buy supplementary audio casettes of singing insects.

Research workers from quite diverse biological disciplines have been interested in acoustic behaviour. The methods of sound production and the acoustic regimes are fascinating because of the variety of mechanisms that have evolved, and the ways in which problems of transmitting acoustic signals, that are peculiar to small animals, have been overcome. Zoologists with a biophysical turn of mind have investigated the transmission and reception of acoustic signals. Because much arthropod sound occurs in a sexual context it can be involved in determining mate choice, sexual selection and sexual isolation. Insect song has been used widely by ethologists and sociobiologists to provide model systems to investigate these concepts. The stereotyped nature of arthropod acoustic communication has made it almost ideal material to use in the rapidly expanding area of neuroethology. Over the last two decades great advances have been made in elucidating the central control of the motor patterns underlying stridulation in grasshoppers and crickets. Probably even more impressive has been the work done on the processing of acoustic information in the central nervous systems of Orthoptera.

Along with the enormous growth of interest in all aspects of sound production and reception in arthropods has arisen an understandable tendency for narrow specialization. This even has some geographical basis. Much of the neurobiology has been carried out in Europe, predominantly in Germany, while the more behavioural and ecological advances have tended to come from North American researchers. I have attempted in this volume to bring together the different approaches and hope that the synthesis will be useful to all.

I have benefited from the advice and support of many friends and colleagues during the preparation of this book. My wife, Dr Leonie Ewing, in particular, spent many hours in reading the manuscript and attempting to eliminate my solecisms, while the neurobiological chapters were scrutinized by Dr Jaleel Miyan, who also helped with the scanning electron micrographs. I am grateful to the following for permission to reproduce copyright material: Academic Press Inc. Ltd.; Akademiai Kiado es Nyomda; American Institute of Physics; Blackwell Scientific Publications Ltd.; Bruel and Kjaer; Company of Biologists Ltd.; E.J. Brill, Leiden; Entomologica Sallskapet; Entomological Society of America; Evolution; Harley Books; Macmillan Magazines Ltd.; Nederlandse Entomologische Vereneging; Neumann–Neudamm, Verlag für Jagd and Natur; Pergamon Press PLC; Psyche; Springer-Verlag, Heidelberg; The Royal Society; The Royal Society of New Zealand; Verlag Paul Parey, Berlin. I also thank the authors for allowing me to use illustrations from their publications: these are cited individually in the legends to the figures. Others who gave help in a variety of ways include Professor M.J. Claridge, Dr Murray Campbell and Mr D.F. Cremer.

A.W.E.
Edinburgh
1989

1. Biophysics of Sound and Vibration

Animals can perceive mechanical stimuli transmitted through the medium of air, water or a variety of solids such as soil, sand and different plant materials; leaves, stems or wood. Both the mechanisms that they utilize to produce mechanical stimuli and the mechanoreceptors that respond to the stimuli have evolved in response to pressures exerted by the properties and constraints of the transmission medium. Some knowledge of the ways in which such stimuli are transmitted is therefore essential if we are to understand acoustical and vibrational communication. Propagation through different fluids is governed by similar principles, while in solids the rules are somewhat different and are both more complex and less well understood.

I have attempted to keep the following descriptions as simple and untechnical as possible, and have only discussed those aspects of the subject which are strictly relevant to arthropod communication. As well as many textbooks on acoustics and vibration there are several good modern reviews on various aspects of bioacoustics to which the reader can refer. I go into more detail in the text when discussing specific examples of sound production and reception, while the Glossary provides some of the relevant formulae.

SOUND TRANSMISSION IN FLUIDS

In studying acoustics, a start is usually made with an idealised, perfect sound source such as a pulsating sphere, called a monopole source, or an oscillating sphere, called a dipole, which act in free space. A biologically more relevant source would be a diaphragm such as a drum skin, although its properties are more complex. If the skin of a drum is struck, causing it to vibrate, one can plot the displacement of the skin against time and relate the motion to different parameters of the sound (Figure 1.1).

As the skin vibrates the molecules of the fluid become alternately compressed and rarefied. These cyclical oscillations become attenuated away from the sound source because of spreading and losses in the form of heat. An idealised sound source would produce a pure tone with a fixed frequency which will depend on the properties of the diaphragm, such as its inertia and tension. One consequence of the molecular motion

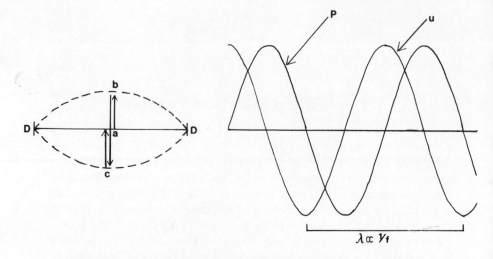

Figure 1.1. If a diaphragm, such as the skin of a drum is made to vibrate, it
will oscillate, passing through the positions a–b–a–c–a, etc., as long as energy
is provided. These oscillations result in sinusoidal fluctuations in pressure (P)
and particle displacement (u) which are 90° out of phase, as shown.
Wavelength (λ) varies inversely with frequency (f).

is a sinusoidal change in pressure about the mean value, and it is this
variation in pressure which produces the sensation of hearing in many
animals, including ourselves.

Frequency and wavelength

The frequency of a sound or vibration is the number of cycles of pressure
change with respect to time for which the SI unit is the Hertz (Hz),
representing the number of cycles per second. Humans can hear frequencies of between 30 Hz and 15 kHz, while insect organs of hearing
cover the range between 15 Hz and 120 kHz, although that of individual
species is often very restricted. The wavelength, that is, the distance
covered by one cycle (λ), is dependent upon the velocity of sound. In air
at sea level this is about 340 metres s^{-1} and thus the wavelength of a
340 Hz note would, for example, be 1 metre, and that of a 3.4 kHz note
would be 10 cm (Figure 1.2). In water the propagation velocity of sound
is around 1450 metres s^{-1} and the wavelengths for the same frequencies
are thus about four times greater than in air.

Pressure and particle velocity

So far we have only considered cyclical pressure changes, but accom-
panying these are the movements of the molecules of the medium which
are being successively accelerated and decelerated. The power that is

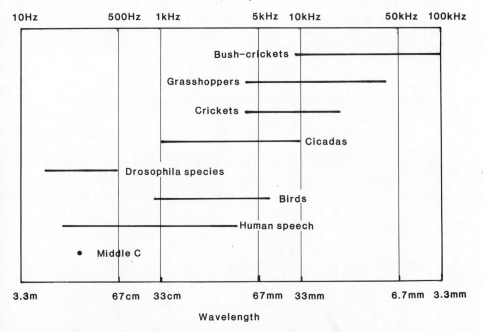

Figure 1.2. The relationship between frequency and wavelength. The ranges are given for a variety of biological signals. Within groups of insects and birds the ranges for individual species are likely to be considerably smaller than those shown.

produced is the product of pressure and particle velocity. It is important to distinguish between these two components of an acoustic signal, because, while the majority of vertebrates possess pressure receptors, this is not true of many arthropods. One important consideration is that while pressure is a scalar quantity, particle displacement has directionality, and thus a receptor which responds to the latter has inherent directional properties.

The second consideration arises from the fact that while pressure and particle displacement are in phase at a distance from the sound source, as the source is approached, a phase angle of 90° develops between them. If we consider a sound source which is small relative to the wavelength of the sound that is being produced, and this is a common occurrence in insects, then close to the source the energy due to the particle velocity is much greater than that of the sound pressure. However, pressure attenuates less with distance than does particle velocity, and there is thus a crossover point: the region where particle displacement is larger is called the near field, and where pressure is greater, the far field. The boundary between the two is difficult to determine because it depends on

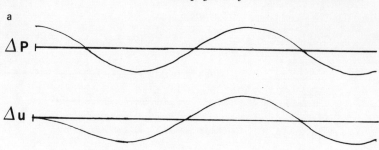

a

Δ P

Δ u

Distance from sound source

b

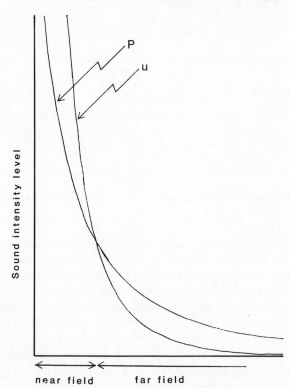

near field   far field

Distance from sound source

Figure 1.3a. The phase relationship between pressure and particle movement close to a spherical sound source. To simplify the picture the effect of attenuation has been removed and the change in velocity (Δu) is plotted against change in pressure (ΔP). Close to the source the two parameters are 90° out of phase, but one wavelength from the source they are in phase and act as a plane wave.

Figure 1.3b. Attenuation of pressure and particle displacement. The crossover point marks the boundary between the near and far fields.

Table 1.1. The sound pressure levels (SPL) in air achieved by some common insects in relation to the echolocating cries of bats, which are probably the most intense sounds produced by animals, and some human activities. The reference level is usually taken as $20\,\mu$Pa $(= 2 \times 10^5 \text{ Nm}^{-2})$.

Sound Pressure level	Decibel level	Signal
200 Pa	140	Human pain threshold
	125	Aircraft jet engine Vespertilionid bat
20 Pa	120	
	115	Mole-cricket
	105	Bush-cricket
2 Pa	100	
	95	Field cricket
0.2 Pa	80	
	75	Human speech
0.2 Pa	60	
	50	Drosophila
0.002 Pa	40	
0.0002 Pa	20	Quiet whisper
20.0 μPa	0	Threshold of human hearing

a number of factors, some of which are not easily characterized, such as the nature of the sound source itself (Figure 1.3a, b).

For most biological sound sources the boundary will be considerably less than one wavelength, and consequently receptors which respond to particle displacement will be efficient for close-range communication, while pressure receptors will be of use at a distance. One should note that in water, for a given frequency, the near field–far field boundary will be about four times farther from the source than it would be in air.

Amplitude

From our simple model of a vibrating diaphragm, it is obvious that the larger the amplitude of the movement, the louder will be the sound produced. Sound levels are normally expressed in decibels (dB) and it is important to be clear from the outset that the decibel is not an absolute but a relative measure of loudness and must always be related to a reference level. We are usually concerned with sound pressure levels (SPL) and

$$\text{SPL} = 20 \log_{10} p/p_r \,\text{dB}$$

where the measured level is p and the reference level p_r. The latter is normally $2 \times 10^{-5} \text{ Nm}^{-2}$ which is a level of sound near the threshold of human hearing at its most sensitive frequency of 4 kHz. One should note that other reference levels are sometimes used, but any publication on acoustic behaviour normally specifies the reference used. Table 1.1 provides an idea of the SPLs achieved by members of some of the insect groups dealt with in this book and relates them to our everyday experience.

Although less usual, it is sometimes convenient to describe loudness in terms of particle velocity where the zero dB level equivalent to the SPL reference is $0.5 \times 10^{-7} \mathrm{ms}^{-1}$.

Sound intensity (I) is the rate of energy transfer at a given point and is, as mentioned above, the product of pressure and particle velocity. The intensity varies with the square of sound pressure, and thus the intensity level of a sound, $\mathrm{IL} = 10 \log_{10} \mathrm{I}/\mathrm{I_r} \, \mathrm{dB}$. Intensity is measured in Wm^{-2} and $10^{-12} \mathrm{Wm}^{-2}$ is the O dB reference level normally used for Ir. Decibels provide a logarithmic scale and Table 1.1 shows the relationship between decibels and sound pressure (or particle velocity) for some common sounds and biological signals. Note that decibels are often used to compare two different sound levels. For example, it is quite valid to state that one sound is 20 dB louder ($+20$ dB) than another, meaning that the sound pressure level is 10 times greater, or that it is 6 dB quieter (-6 dB), that is, having half the pressure.

THE SOUND SOURCE

Because particle velocity and pressure are out of phase in the near field, the load on the sound source is reactive, but for high acoustic efficiency the load must be resistive, that is, the two components must be in phase. Where the sound source is small in relation to the wavelength of the sound produced, the air close to it behaves reactively, and efficiency is therefore low, and it is not until the diameter of the source approaches one wavelength that high acoustic efficiency is obtained. This has important consequences for insects, many of which are small but nevertheless produce relatively low-frequency sounds. Such insects can only communicate in the near field, where particle velocity is high, and have to use receptors sensitive to this component. In order to be acoustically efficient, an insect must be large or produce high-frequency sounds or resort to a variety of behavioural and physiological stratagems. One of the most interesting aspects of insect bioacoustics is how mechanisms have evolved to overcome the physical limitations of small size.

The directionality of a sound is dependent upon the type of source. Thus, a point source or monopole in free space will produce spherical spreading, a dipole produces a figure of eight pattern and a piston or diaphragm, a spreading beam. The detailed pattern in other than a monopole can be complex, and once again depends on the size of the source relative to the wavelength of the sound (Figure 1.4).

Biological sound sources may approximate to idealised ones but more

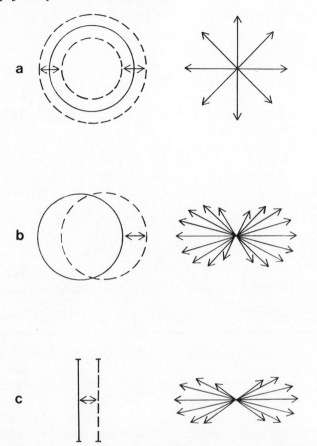

Figure 1.4. The sound radiation patterns for different sound sources. a, Monopole. A pulsating sphere gives rise to spherical spreading of sound waves; b, Dipole. An oscillating sphere produces a cardioid pattern; c, Vibrating piston. A Bilobed pattern of sound radiation is found whose details depend upon various factors such as wavelength and properties of the piston.

All of these patterns are valid only for the far field and where the wavelength of the sound is large in relation to the diameter of the sound source. This is true for the majority of insects excepting some which have ultrasonic songs.

often have features of more than one type and, moreover, they never function in free space. The existence of barriers which reflect and absorb sound has a large effect on its directionality. This is particularly true of sound in water where the difference in density between air and water causes an impedance mismatch and most of the sound will be reflected at the interface at the surface.

A sound-radiating structure, such as a fixed string, diaphragm or

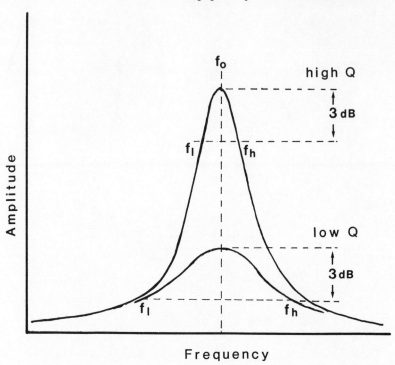

Figure 1.5. The characteristics of high and low Q signals. f_h and f_l are the high and low frequencies respectively at which the signal amplitude is $1/\sqrt{2}$ (-3 dB) that of the peak signal, f_o. $Q = f_o f_h - f_l$.

piston will, if vibrated, produce a note whose frequency will depend primarily upon its physical dimensions and its tension. This is the fundamental or first harmonic. It will, in addition, produce a series of overtones which, in the ideal case of a fixed string will be harmonics or integral multiples of the fundamental, i.e., $\lambda/2$, $\lambda/3$, ... λ/n. For a structure other than a fixed string, or where the parameters are other than ideal, the relationship between the fundamental and the overtones does not conform to a harmonic series. Such sounds are more difficult both to characterize and to predict with regard to the way in which they will behave.

In spite of the above strictures it is interesting to observe that many insects do produce notes which are relatively pure. One way in which they manage this is to produce sounds near the resonant frequency of the sound-radiating surface. This is the frequency at which the amplitude of the vibrations is at a maximum and the mechanical impedance is at a minimum. Under these circumstances not only are pure notes produced but the efficiency of sound production is high. The purity of a note is described by its Q value (quality factor) which is measured as the

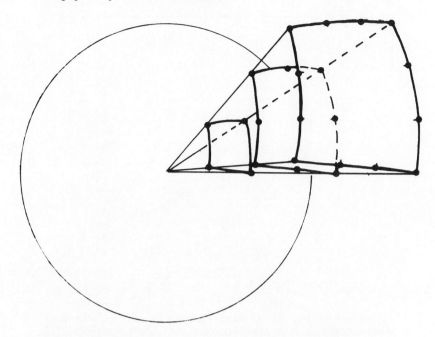

Figure 1.6. Spherical spreading of a sound wave from a point source to demonstrate why attenuation of sound pressure follows the inverse square law.

resonant frequency divided by the frequency bandwidth at the -3 dB points (Figure 1.5)(Michelsen and Nocke 1977; Michelsen 1983).

Attenuation

Sound pressure level decreases with distance from a sound source. The most obvious reason for this is simply the geometrical spreading of the sound wave. In free space and under far field conditions, sound intensity obeys the inverse square law, and the pressure therefore halves for each doubling of the distance ($I \propto p^2$), i.e., changes by -6 dB (Figure 1.6). In the near field the attenuation will be more rapid. However, animals do not exist in free space and the ideal situation is greatly modified by a large number of environmental factors (Michelsen 1985).

Depending upon the causes of energy loss, the type of sound source and geometry of the sound field, attenuation will normally be much greater than -6 dB/dd (dd = distance doubled). On the other hand, in certain circumstances, where the sound is channelled in some way, the loss can actually be less than this. For example, in the sea low-frequency sound waves are trapped within the thermocline and the spreading is therefore cylindrical rather than spherical.

The sources of sound dissipation can be placed in two broad categories, losses internal to the medium and those occurring at its boun-

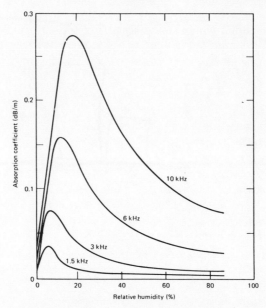

Figure 1.7. Adsorption in air of sound of different frequencies as a function
of relative humidity. Small amounts of water vapour have large effects,
particularly at lower frequencies. The graph also demonstrates the overall
increase in attenuation with frequency. (From Kinsler *et al.* Fundamentals of
Acoustics, John Wiley & Sons, 1982).

daries. Sound dissipation due to absorption within the medium is a
difficult and incompletely understood phenomenon, but the factors that
influence it are known. These are the constituents of the medium, the
temperature and the wavelength of the sound. At audio frequencies the
most important constituent is water vapour, and absorption increases
with relative humidity, the effect being most marked at higher sound
frequencies as illustrated in Figure 1.7. Under extreme conditions the
sound attenuation can be several dB/m.

Inhomogeneities within the medium can also cause sound losses due
to scattering of the sound waves. In air this may be caused by smoke or
fog particles and once again the effect is greatest at the higher frequencies. Many inhabitants of large cities subject to atmospheric pollution
will be aware of the sound-deadening properties of fog.

In water, where losses due to absorption are much less important that
in air, scattering due to air bubbles resulting from wave action can cause
considerable sound attenuation. In shallow ponds the attenuation of
sounds in the range 2–12 kHz was found to be roughly − 6 dB/dd away
from the shore. However, during periods of intense sunlight, photosyn-
thetic activity of water plants causes production of gas bubbles which
results in a dramatic attenuation of sound waves which varies with
frequency (Theiss and Prager 1984).

The remaining sources of sound loss are at the boundaries of the transmitting medium where sound scattering and absorption occur. In the natural world animals are usually communicating acoustically in or around a variety of objects such as grass, trees, shrubs and, of course, the ground. Depending upon the physical nature of the object and the angle at which the sound wave impinges on it, a proportion of the energy will either be absorbed or reflected. A complex condition may exist where the original sound wave and a reflected component of it meet and interact. Where the constituent waves are in phase the interaction will be constructive, and where out of phase, destructive, causing irregularities in the degree of attenuation as one moves away from the source.

Finally, there are possible effects resulting from sound refraction. These are due to variations in the speed of sound which occur because of factors such as differential temperature, wind speed or humidity in air or in water, temperature, salinity and pressure. Gradients or discontinuities in these environmental factors can cause a bending of sound waves resulting in either sound 'shadows' or in sound 'channels' which, for example, occur in the sea as described above.

The attenuation of sound in field conditions is affected by so many factors which are ill-understood and so difficult to measure with accuracy that it is almost impossible to make useful theoretical predictions. Nevertheless, an awareness of the factors involved and a knowledge of their possible magnitudes are clearly essential in interpreting the ecological and evolutionary adaptations of sound-producing animals (Aiken 1985, Michelsen 1978, Michelsen and Larsen 1983).

TRANSMISSION OF VIBRATIONS

Vibration in solids

Vibrations are transmitted through solids by a variety of wave forms. Some of these have been demonstrated to be of biological interest in that they are used in communication or are detected for some other purpose such as prey location. Wave propagation is more complex in solids than in air or water, partly because of the different wave types that occur, and because some solids, such as plant stems or sand, have different transmission properties. Thus the transmission velocities and rates of attenuation can differ hugely and one can observe complex patterns of selective frequency filtering (Figure 1.8). However, in many circumstances, transmission through the substrate can be a very efficient method of communication, particularly for small animals, due to lower attenuation of mechanical energy compared with air.

Most animals live on the surface of a solid, that is, on the interface between a solid and air, or a solid and water, while a few live on the interface between two fluids, water and air. What we will be concerned with mainly are the different wave types and the distortions that occur

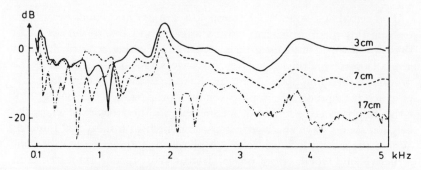

Figure 1.8. The attenuation of sinusoidal vibrations along a thin (*c.* 1 mm
diameter) plant stem. At below 2.5 kHz complex frequency filtering is seen
and vibration amplitudes can change by up to 20 dB with very small
alterations in frequency. (From Michelsen, Fink, Gogala and Traue 1982).

at these boundaries and as a result of the interactions between the various
wave forms.

The simplest and most obvious of these are longitudinal waves which
are analogous to those occurring in fluids (Figure 1.9). It is unlikely that
these could be utilized for communication except perhaps in the rare
situation where an animal was entirely surrounded by and in contact
with the substrate, as in a wood-boring insect. Such waves, however,
have an effect at the surface, as the longitudinal compressions and
extensions will cause corresponding lateral expansion and contractions.
These quasi-longitudinal waves are smaller than the pure longitudinal
waves by a factor of between 10^{-2} and 10^{-3}, but are nevertheless capable
of stimulating vibration receptors of animals standing on the surface.

Transverse waves are those in which the particle movement is perpen-
dicular to the direction of wave motion, in contrast to longitudinal
waves, where it is in line with the direction of travel. As transverse waves
are in plane with the surface it is again unlikely they are perceived by
animals except in the special case of the vibrations occurring in the webs
of some spiders.

One type of wave which has been demonstrated to be important in
communication is the bending wave which occurs in long thin structures
such as plant stems. Propagation velocity of bending waves depends
upon the physical properties and dimensions of the medium and also on
the wavelength of the vibration. This has the important consequence for
communication that vibrations of different frequencies travel away from
the sound source at different speeds, and are therefore said to be dispersive. Also, for such waves, if the vibrations are produced as pulses, which
is usual in communication, then the velocity of the pulses, the group
velocity, is twice that of the phase velocity (Figure 1.10).

Most of the foregoing descriptions are relevant to propagation along
structures where the wavelength of the vibration is considerably less

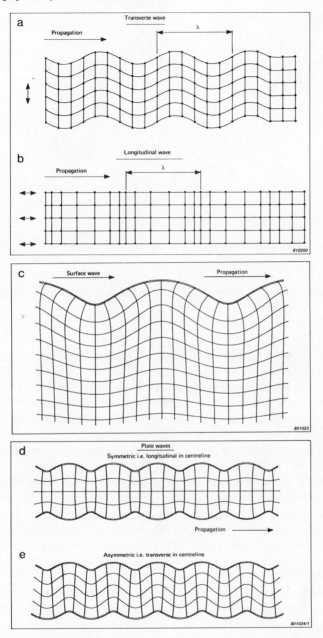

Figure 1.9. The different types of waves which can be propagated in solids. Transverse (a) and longitudinal (b) waves can interact at the surface of a solid to produce Rayleagh waves (c). Symmetric (d) and asymmetric (e) waves are found at the surfaces of thin plates of material. (From Rindorf 1981).

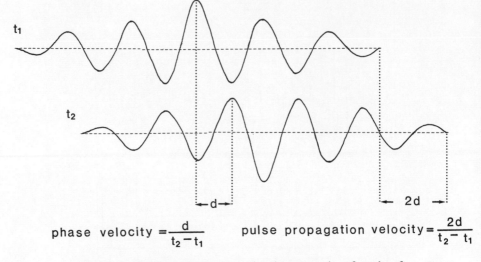

$$\text{phase velocity} = \frac{d}{t_2 - t_1} \qquad \text{pulse propagation velocity} = \frac{2d}{t_2 - t_1}$$

Figure 1.10. The figure demonstrates that the propagation of a pulse of bending waves in a solid, that is, the group velocity, is twice that of the phase velocity.

than the cross-section of the structures involved – these are beams and thin plates. However, on a substrate such as the ground or a tree trunk, this may not be true, and here we find a variety of surface waves which apparently result from the interactions of longitudinal and transverse waves. These are named Lamb waves, Love waves and Rayleigh waves, and the last of these at least have been demonstrated to be used by animals; for example, scorpions use them to locate prey in or on the sand (Brownell 1977, Markl 1983, Gogala 1985).

It should be emphasized that any mechanical stimulus will simultaneously produce several of the wave forms described above, but only in a few cases has an analysis been carried out to identify which are biologically useful. A knowledge of the physical properties of waves will give some idea as to what kind of information they can transmit.

Water–surface vibration

Surface waves of a type different to those which are found in solids occur in water. Their properties depend upon the depth of water relative to wavelength, and in deeper water where depth is greater than $\lambda/6$, concentric waves are produced which, like the bending waves in solids described above are propagated dispersively. In shallower water however, non-dispersive elliptical waves are produced. All surface waves are rapidly attenuated, and, as in air, the effect is greatest at higher frequencies. This means that communication using water surface waves is likely to be restricted to the use of low frequencies and to be effective over a relatively short range.

SUMMARY

Because of their small size most arthropods face a number of problems with regard to the production of airborne sound. For efficient sound production the structures which radiate sound energy must not be too small compared with the wavelength of the sounds produced. This means that small insects should sing in the kilohertz range and many insects indeed do so. However, sound attenuation increases with increasing frequency, and there must therefore be a trade-off between these two competing factors. One way of overcoming this problem is to communicate at close range. While the sound pressure levels produced at low frequencies are small, close to the source the particle velocity may be extremely high. The latter attenuates much more rapidly than sound pressure, but given the appropriate sensory receptors, small insects can communicate efficiently over short distances using low-frequency sounds. Alternatively, because attenuation through many solid structures is much less than in air, insects and other arthropods can communicate using substrate-borne vibrations. A good example of this is provided by those Hemiptera which use abdominal tymbals in sound production. While the larger species, the cicadas, produce intense airborne sound, the small planthoppers transmit their signals along plant tissues.

2. Mechanisms of Sound Production

The first land animals to use sound for communication were probably the arthropods. Stridulatory structures have been identified in the fossil scorpion, *Eubuthus*, from the Carboniferous Era, and are found widely in Triasic genera (Willis 1946). If the sounds were used for defence as in the modern scorpions, then there must have been other arthropods present which were able to perceive the signals (Alexander 1958). Until the advent of the modern orders of insects in the late Permian to early Triassic, around 250 million years ago, the majority of animal sounds would have been inadvertent results of activities such as location or feeding. All acoustic signals must have evolved adventitiously from behaviours originally unconnected with communication. In most terrestrial vertebrates the activity which has given rise to acoustic communication is breathing, with vocal cords evolving independently in mammals, birds and amphibia. As the rate of breathing is affected by emotional state it is easy to envisage how sounds produced as a by-product of respiration came to have communicatory functions. We acknowledge this in human beings when we talk of 'gasps of surprise' or 'screams of delight'.

In arthropods there appears to be no connection between the evolution of acoustic communication and emotional state for the obvious reason that their respiratory systems are quite different from those of vertebrates. The most important factor has been the possession of a hard exoskeleton where almost any movement of the body parts will produce sounds or substrate vibrations. One consequence of this difference in the evolutionary origin of acoustic behaviour is that the variety of sound-producing mechanisms found in the arthropods is very much greater than in vertebrates. Although sounds are produced in the latter by means other than vocalisation, such as hand clapping in humans, drumming in snipe, leg thumping in rabbits or the rattle of the rattlesnake, these are of relatively minor importance.

There is no entirely satisfactory classification of the diverse methods of sound production in arthropods, and I have made use of five broad categories which encompass most of the known examples. These are stridulation, percussion, vibration, click mechanisms and air expulsion. This classification is solely mechanistic and does not take into account the methods of transmission. The first four classes could function equally in air, water or a solid. Indeed in many cases the sounds or vibrations produced by arthropods are propagated simultaneously through more than one medium, and it is sometimes difficult to establish the precise channel of communication. This is one of the reasons for not using a classification based on the transmission medium.

PERCUSSION

Percussive mechanisms are those in which two structures are brought forcibly together. The efficacy of this method for producing sounds or vibrations in arthropods is due to the possession of an exoskeleton, and there are many examples where either two body parts are clapped or, more commonly, some hard skeletal structure is struck against the substrate.

Percussion of body parts

One reasonably well-understood example of a percussive mechanism involving two body parts is found in members of the Australian moth genus, *Hecatesia* (Bailey 1978). In these moths an area of the costae of the fore wings is modified to form small hard knobs which have been called castanets. These are repeatedly struck together at the top of the wing stroke to produce sounds which have given these insects the popular name of 'whistling moths'. Two of the three known species have been looked at, and of these, one, *H. exultans*, produces the sounds while at rest, while the other, *H. thyridion*, does so during a 'nuptial' flight. The sounds are produced by males only and are presumed to have the function of attracting potential mates. Both species produce trains of pulses at about $200 \, s^{-1}$ in *H. exultans* and $80 \, s^{-1}$ in *H. thyridion*. Each pulse in the former species consists of an oscillation which is lightly damped, suggesting that the sound-radiating surfaces are uncoupled from the main mass of the moth. Thin crescentic areas of clear corrugated cuticle posterior to the castenets probably have this function, the wings distal to the castanets being set into resonance (Figure 2.1).

A power spectrum of the pulses shows the maximum energy to be fairly sharply peaked at between 25 and 30 kHz in *H. exultans*. The shape of the pulses with a gradual increase of amplitude over several cycles indicates that we are dealing with a resonant structure which is being loaded. This could be the space contained by the wings, which approximates to a sphere of radius 5–6 mm, that is, half the wavelength of the carrier frequency of pulses.

In *H. thyridion* the individual pulses are shorter than in *H. exultans*. This is due to the wings being opened after each collision which has the effect of dissipating the energy. The view that the wings radiate the sound energy is reinforced by the observation that the fundamental

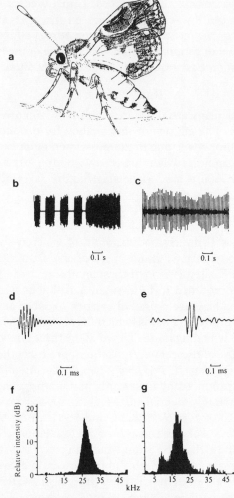

Figure 2.1. Percussive song in the moth *Hecatesia*. a, *H. thyridion*, showing the 'castanets' on the fore wings; b, call of *H. exultans*; c, Call of *H. thyridion*; d,e, Wave forms of individual pulses in *H. exultans* and *H. thyridion*; f,g, Frequency spectra of the calls of *H. exultans* and *H. thyridion*. (a. From a photograph by P. Merlin Crossley; b–g. From Bailey, 1978).

frequency of the pulses in *H. thyridion* is lower at between 17 and 20 kHz and the wings are correspondingly larger.

This relatively simple system demonstrates the interrelationship between the behaviour of the moths, the specialization in the morphology of the wings and some of the acoustic principles described in the previous chapter.

While the carrier frequency of *Hecatesia* sound reaches into the

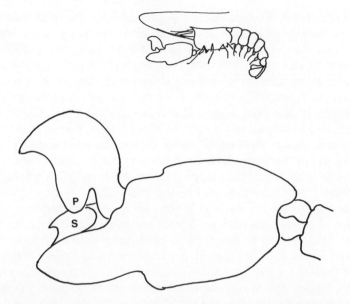

Figure 2.2. The enlarged chela of a pistol shrimp, *Alpheus*, illustrating the plunger (P) on the dactylopodite and the socket (S) on the propodite into which it engages to produce the loud sounds that give these crustaceans their popular name.

audible range for humans, at least one other species of moth, *Heliotis zea*, which does not have any specialized sound-producing apparatus, produces ultrasonic pulses during flight by clapping its wings together. It is not unlikely therefore that this mechanism is more widespread that has hitherto been suspected (Agee 1971). The use of 'bat detectors' designed to study bat echolocation may reveal more examples.

Other examples of percussive signals produced by bringing body parts together are found in the Pistol Shrimps, family Alpheidae, (Johnson, Alton Everest and Young 1947), in some Spur-throated Grasshoppers (Acrididae: Cyrtacanthacridinae) which snap their mandibles (Alexander 1960) and in the brachypterous grasshoppers of the family Pamphaginae which tap their wings against the femorae (Dumortier 1971).

The first of these is of interest because, although the structural modifications for sound production and their probable function have been known since the last century, it is only since the Second World War that the extent to which the activities of species of *Alpheus* and *Synalpheus* contribute to underwater noise has been realised. With the advent of underwater listening devices related to submarine warfare it became evident that there was a considerable amount of previously unsuspected acoustic activity of biological origin. In shallow waters the noise levels can be 30 dB above ambient due to the snapping of Pistol Shrimps.

The sounds are produced by one greatly enlarged and modified chela, of which the moveable dactylopodite has a protusion or plunger that fits a socket on the propodite. In some species there is a sucker which holds the dactylopodite in the fully open position, and one imagines that when muscular contraction overcomes its suction the closing action of the claw is enhanced. A loud click is heard as water is forced out of the socket by the plunger. It has been suggested that it is the water jet which is used for defence and offence and the sound is incidental; however, the almost continuous chorus that exists makes this explanation unlikely. The shrimps live in aggregations either in burrows in the substrate or in coral fissures. It is likely that the snapping sounds are involved in spacing and also possibly to attract potential mates (Figure 2.2).

Observations by Hazlett and Wynn (1962) on the behaviour of the shrimps show that the enlarged chelae are seldom used for feeding or against heterospecifics. Snapping is performed almost exclusively in response to conspecifics and in particular towards males. While the earlier work quoted above found considerable energy in the snaps up to 12 kHz, Hazlett and Wynn show that the sound pulses have a duration of around 45 ms with only the initial 3 ms containing high-frequency components while the remainder of the pulse was comprised of a 75–150 Hz component.

Percussion involving the substrate

In contrast to the relatively small number of examples that have been described where different body parts are struck together, percussive signals resulting from contact with the substrate are many. Examples are to be found among ants, termites, book lice, stoneflies, beetles, bugs, grasshoppers, spiders and crabs.

In the majority of cases the structures involved in sound production are relatively simple and specialized mechanisms may be entirely lacking. One reason for this lack of sophistication is that the signals produced are almost invariably transmitted through the substrate. Because of dispersive propagation, the spectral composition of such signals is probably relatively unimportant for communication in most cases. Complex sound-producing mechanisms are more usually asso-ciated with the production of signals with a specific frequency structure.

The sounds that are produced are often described as 'drumming'. In book lice or stoneflies the abdomen is tapped against the substrate (Pearman 1928, Rupprecht 1976), crabs and Orthoptera use their appen-dages (Salmon and Horch 1972, Sismondo 1980) while termites and beetles make use of their heads. Specialized 'hammers' are found on the ventral apical region of the abdomens of some stoneflies, spiders and book lice (Uetz and Stratton 1982). In the latter it appears, atypically, that the signals are produced by females, while in the stoneflies sounds

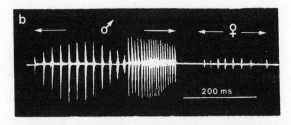

Figure 2.3. Percussion in stoneflies. a, *Cannia bifrons* tapping the substrate with the end of its abdomen; b, An oscillogram of the vibrations produced by the male *Isoperla phalerata* and the female response. (From Szczytko and Stewart 1979).

are produced by both sexes. Complex, species-specific patterns of pulses may be produced by calling males which are answered by virgin females (Figure 2.3).

The structures associated with sound production in the stoneflies are located variously on the 7th to 9th abdominal segments and consist of a peg or hammer. Some species possess a moveable vesicle bearing tactile hairs, or the hairs alone, which provide sensory feedback enabling the animal to position the abdomen relative to the substrate (Rupprecht 1976).

Social insects such as termites and ants drum on the substrate to produce warning signals to alert colony members. Thus, workers of the termite, *Zootermopsis angusticollis*, tap with their heads on the floor and roof of their galleries (Howse 1964). A more detailed recent study on the Carpenter Ant, *Camponotus herculeanum*, has revealed similar behaviour, with the vibrations being produced by drumming the head and gaster against the substrate. These ants live in tree-trunks from which the soft spring wood has been eaten out, leaving thin, lignified lamellae. The bioacoustics of this system have been explored by Fuchs (1976) who has shown that the frequency and intensity of the signals depend upon the dimensions of the lamellae. On the thinnest of these, $c.$ 0.5 mm, acceleration velocities of up to $500 \, \text{cm} \, \text{s}^{-2}$ can be measured over a range of frequencies between 200 and 1500 Hz. The attenuation is as little as $0.4 \, \text{dB} \, \text{m}^{-1}$ and, as the lowest response threshold of the ants is under $5 \, \text{cm} \, \text{s}^{-1}$, this gives a maximum theoretical range of the signals of up to 90 cm, although 20–30 cm is a more likely figure. As the substrate becomes thicker, the acceleration velocity decreases and the frequency spectrum is shifted upwards away from the most sensitive range for the ants' receptors of 250–750 Hz. Thus the alarm signals are particularly suited to transmission through the nest rather than on the outside trunk.

Littoral crabs of the family Ocypodidae produce percussive signals which can be detected over a range of several metres. They drum on the ground with either their chelae or ventral surface of the carapace, usually at their burrows' entrances but also underground. Bursts of pulses are produced at a rate characteristic of the species, e.g., $10–20 \, s^{-1}$ for *Uca pugilator*. These produce wide-band frequency vibrations containing energy up to 12 kHz, although much of it is concentrated below 5 kHz. Most of the high-frequency components are attenuated rapidly and filtered out by the substrate of wet sand or mud, and the crabs are particularly responsive to the low-frequency vibrations.

Many spiders, particularly web-building ones, use substrate vibrations in order to locate and capture prey, and it is therefore not surprising that such vibrations are also used in communication. These can be produced by a variety of mechanisms, including percussive ones. Lycosids such as *Lycosa gulosa* may 'strum' with their palps and drum with their abdomens (Harrison 1969), the former sounds comprising pulses with a frequency between 2 and 4 kHz and the latter from 4 to 7 kHz (recorded with a microphone and not via the substrate). While these signals are easily audible to humans, spiders presumably perceive them as ground-borne vibrations with their lyriform organs. However, the palps of many lycosid spiders have stridulatory organs, and the sounds produced by palpal 'drumming' and described as such in the literature could, in some cases at least, be due to stridulation. While spiders appear to possess few specialized structures for percussive signalling, one species, *Hygrolycosa rubrofasciata*, is reported to have a sclerotized abdominal plate for this purpose (Weygoldt 1977).

In spite of the widespread occurrence of substrate-borne percussive signals they have received much less attention than other, more elaborate methods of acoustic signalling. One possible reason is that for such signals to be recorded satisfactorily some type of vibratometer is necessary, and these are less readily available and are more expensive than microphones for recording airborne sound. The latter can be used for recording substrate vibrations, and are often adequate for providing information on the temporal characteristics of signals. Frequency measurements are, however, misleading.

AIR EXPULSION

There are rather few examples of arthropod sound production using the methods widespread in vertebrates, that is, the movement of air through a tube. Such sound are commonly associated with the expulsion of

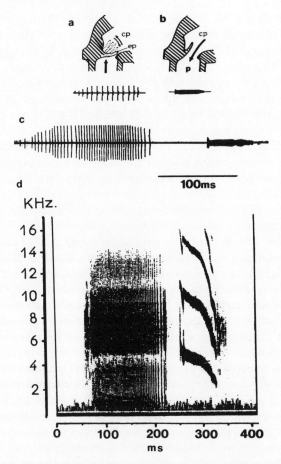

Figure 2.4. Sound production in the Death's Head Hawk Moth, *Acherontia
atrops*. a,b, Longitudinal section of the epipharynx showing how pulsed
sounds are produced during inspiration and a continuous note on expiration;
c, Oscillograms; d, Sonograms of the two sounds. ep = epipharynx;
cp = pharyngeal cavity; p = proboscis. (From Busnel and Dumortier 1960).

unpleasant or noxious substances through the spiracles, as in some tiger
moths (*Rhodogastria* species) and the defensive glands of bombardier
beetles (*Brachinus* species). There is no evidence, however, that the
sounds are other than the by-products of the defensive behaviour or that
they are used in communication.

One example has nevertheless received considerable attention, as it
involves one of Europe's largest and most spectacular insects, the
Death's Head Hawk Moth, *Acherontia atrops*. The source of the sounds
that it produces was first correctly identified by Prell in 1920, who
exhaustively reviews the earlier literature and quotes over 60 references
on the topic, starting with Réaumur in 1734. The sounds had been

variously attributed to movements of abdomen, thorax, wings and mouthparts, as well as to air expulsion through the spiracles and proboscis. Prell implicated forced-air movement through the epipharynx. Busnel and Dumortier (1960) showed that air was drawn into the pharynx through the constriction at the epipharynx, whose movements modulate air flow to produce up to 50 pulses at a rate of 280^{-s}. Loud sounds with a dominant frequency of around 7–8 kHz are produced, while subsequent air expulsion results in a short pulse with a frequency of 9–10 kHz (Figure 2.4). Sales and Pye (1974) report that some African hawk moths in the genera *Nephele* and *Coelonia* can produce sound similar to those of *A. atrops*, and they may therefore possess similar mechanisms.

Cockroaches of the genus *Gromphadorhina* which are found in Madagascar, hiss through modified abdominal spiracles in response to disturbance and also during courtship and aggression. Part of the tracheal system is modified for sound production with the fourth abdominal spiracle having an enlarged opening and leading to an elongated, horn-shaped trachea. This connects with the longitudinal tracheal trunk system through a constriction. Sound is produced when the spiracles, other than number four, close, the expiratory muscles contract and air is forced past the constriction into the horn and so through the enlarged fourth spiracle to the exterior.

The sounds so produced vary, depending on behavioural context, and wide-band hisses containing energy up to 40 kHz with a maximum at around 10 kHz are used when the cockroach is disturbed. In one species, *G. chophardi*, a frequency-modulated song is used in courtship (Fraser and Nelson 1982). While this system has not been subjected to rigorous acoustic analysis, it is obvious that the ability of individual species to produce different types of sound presupposes complex motor control. Nelson (1979) suggests that the enlarged trachea acts as an exponential horn with a resonant frequency of 8.5 kHz while the atypically elaborate spiracular musclature is used to control air flow.

A final intriguing possibility of tracheal sound production is in tsetse flies which make a variety of sounds associated with feeding and mating. These contain frequency components in the ultrasonic range up to 80 kHz (Saini 1985). No altogether satisfactory mechanism has been proposed, but Anderson (1978) has shown that the tracheae supplying the flight muscles are encased in spiral muscle fibres which, on contraction, would expel air from the trachea. He suggests that it is this which is responsible for sound production.

VIBRATION

All sound production must result from the vibration of some structure. However, in this section I will only consider those mechanisms in which body parts are vibrated by direct muscle action without the intervention of some method of frequency multiplication. Stridulatory mechanisms are therefore excluded, to be dealt with in a later section. In the arthropods, sound may be produced by the vibration of almost any area of the body: the wings, appendages, mouthparts and different body regions have all been implicated, although specialized structures are relatively uncommon. As in the majority of percussive signals, the carrier frequency of the vibrations tends to be low and the signals are usually transmitted through the substrate. If in air, they are not usually perceived as pressure changes but in the near field as variations in particle velocity.

Vibration of the wings

Sound is produced as an incidental consequence of flight in insects, and many species have characteristic flight tones which are easily recognized. There is no doubt when one is suffering the unwelcome attentions of a mosquito or tabanid. Observations on horses and cattle also make it clear that they can discriminate biting insects from their flight tones. It is not surprising then that the sounds produced by wing movements, which are probably derived from flight, have taken on an intraspecific communicatory role.

One of the best-known and earliest examples of the flight tone of an insect used in communication is in the Mosquito, *Aedes aegypti*. In common with many diptera, sexually receptive females form a mating swarm over some marker which is often a conspicuous environmental feature. As the flight tone provides a near field signal it cannot be used to attract the male from a long distance. Males are presumably attracted to the swarm by visual cues, either by the swarm itself or by the same landmarks which were used initially by the female.

The co-evolution of sexual behaviour and the development of the flight system are interesting in that on eclosion, the flight tone of the female, that is, the wing beat frequency, is around 400 Hz. This signal is not attractive to males, indeed it is probably beyond their frequency range for hearing, and it is not until 48 hours after eclosion, when the wing beat frequency has increased to 500 Hz, that males respond to the flight tone. This coincides with the onset of sexual maturity in the females. As there is a degree of species specificity in flight tone, this simple mechanism helps to ensure the association of conspecifics in an appropriate state of sexual maturity (Roth 1948).

The best studied example of near field sound communication using wing vibration is in fruit flies of the genus *Drosophila*. Song production appears to be an almost universal trait within this genus and in the related genus, *Zaprionus*. It occurs in both males and females, in sexual and territorial contexts. There is considerable variation in the patterns of song that are found in the different species and also in the form of the wing movements that produce them. In spite of this apparent complexity the mechanism of sound production is relatively simple and the variety is due to the underlying neuromuscular control.

Typically there are two different song types: pulse trains, consisting of a series of regularly spaced pulses, and tone bursts. The former can be made up from single cycle pulses or the pulses may be polycyclic. Tone bursts are continuous sounds with a relatively narrow carrier frequency. They result from low-amplitude wing beats produced by the indirect flight muscles and have carrier frequencies of between 100 and 450 Hz. The frequency appears to be correlated with the degree of wing exten- sion, and those species such as *D. erecta* which sing with their wings extended between 20 and 40°, produce higher-frequency tone bursts than those like *D. melanogaster* which extends one wing to 90°. The same correlation holds for the pulses within the pulse trains, and it is probably a consequence of increased inertial damping as the wing is extended. The wing vibrations that produce pulses are caused both by contractions of the indirect flight muscles, and of the axillary muscles, in particular the subalar. The intra-pulse frequencies can be higher, up to 600 Hz, and the amplitude of the pulses greater than in tone bursts. The forego- ing generalisations do not by any means tell the whole story, and some species, such as *D. micromelanica*, produce tone bursts which are both frequency and amplitude modulated while others produce very complex patterns of song. To date the mechanisms of sound production have been studied only in relatively few species (Ewing 1977, 1979a).

The bioacoustics of *Drosophila* song have however been subject to scrutiny by Bennet-Clark (1971), and his study constitutes one of the few attempts to evaluate the role of near field acoustic communication in insects. On the basis of the size of the wing, the amplitude of the wing vibration and the carrier frequency of the song, he calculated that the particle velocity at the antenna of a female (the receptor for song) would be in the region of 95 dB (re. $0.5 \times 10^{-7} \text{ms}^{-1}$). As particle velocity attenuates by the inverse sixth power of distance (18 dB/distance doubled), acoustic communication is only possible over a very short distance. Unfortunately no direct measurements have been made to discover the threshold sensitivity of the antenna of *Drosophila*, but behavioural observation would suggest the range of audibility of the song to be at a maximum of 10 mm. One should note that by comparison with particle velocity the sound pressure level is only 51 dB (re. $2 \times 10^{-5} \text{Nm}^{-2}$), some 158 times quieter (Figure 2.5).

The limitation of acoustic or vibrational communication to the near field should not be seen as necessarily disadvantageous to the male. Rather it ensures privacy of communication, and males can concentrate their courtship on a single female without being heard by either potential

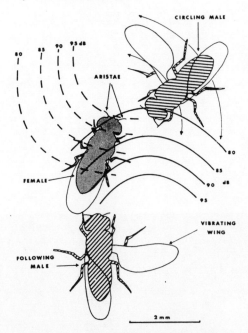

Figure 2.5. Sound particle velocity produced by courting male, *Drosophila melanogaster*. The extreme attenuation illustrated by the contours demonstrates the need of the male to be close to the female in order to be heard by her. OdB reference level = $0.5 \times 10^{-7}\,\mathrm{ms^{-1}}$. (From Bennet-Clark 1971).

predators or other males. The latter could be an important factor, as it is known that male courtship song stimulates males as well as females, and an overheard male song could invite unwanted competition. In some species it is clear that males orientate to the female in such a way that the sound source, the wing, is placed as close to the antenna of the female as possible. In order to achieve this a male will circle in front of the female and extend the wing which is nearest to her head.

The production of vibratory signals like those of drosophilids has been reported from several different groups of Diptera. In particular, because of their economic importance, the behaviour of fruit flies of the family Tephritidae has been looked at, and they have been shown to use songs in courtship and aggression (Keiser, Kobayashi, Chambers and Schneider 1973, Sivinski and Webb 1986, Webb, Burk and Sivinski 1983).

In a manner similar to *Drosophila*, honey bees produce pulse trains by wing vibrations during the waggle dance. The patterning of the sounds is very reminiscent of that found in some *Drosophila* species and has a carrier frequency of 280 Hz. The signals are modulated by movements of the bee's abdomen. Sound pressure level measured with a probe

a

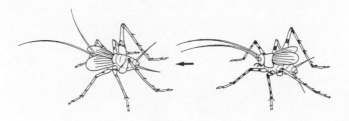

b

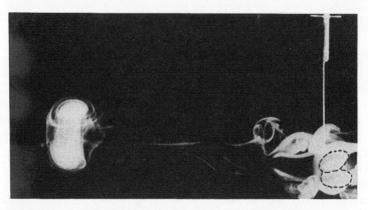

Figure 2.6. a, Wing flick of *Phaeophilacris spectrum*; b, A vortex ring
generated by a simulated wing flick. The position of the wings is indicated
by dashed lines. (From Heinzel and Dambach 1987).

microphone at 1–2 cm from the dancing bee was only 73 dB. However,
the particle velocities at the wing edges are about $1 \, \mathrm{m \, s^{-1}}$ (*c.* 110 dB) and
follower bees appear to position themselves so that they place their
antennae in this region. While the antenna has not been definitely
identified as a sound receptor, it is likely that in bees it does have this
function as it does in many insects. The high rate of attenuation of the
signals ensures that only those bees close to the dancing one will receive
the acoustic message contained in the dance. On a comb one can often
observe bees from different food sources dancing and the near field
nature of the signal will eliminate possible interference between them.

A final interesting example which only marginally fits our definition of
wing vibration is in *Phaeophilacris spectrum*, an African cricket. During
aggression and courtship, males flick their wings forwards over their
heads at a rate of $10 \, \mathrm{s^{-1}}$. These slow flicks produce what has been termed
'infrasound' and contain little in the way of acoustic energy. However,
neurophysiological evidence has shown that female *Phaeophilacris* can

perceive the flicks at a range of 15 cm (Kamper and Dambach 1979). This led Heinzel and Dambach (1987) to suggest that the effective stimulus was not in the form of conventional acoustic waves but travelling air vortices cast off from the tip of the wings at the termination of each flick.

They carried out experiments using an electromechanical transducer to move a pair of wings so as to mimic the actual motion. Using titanium tetrachloride or *Lycopodium* spores to visualize the air movements showed that vortices were produced and that they travelled a distance of up to 15 cm (Figure 2.6). The particle velocities within these vortices were measured by means of stroboscopic examination of individual spores and approached $15 \, \text{cm} \, \text{s}^{-1}$, a figure likely to be considerably in excess of the receptor thresholds.

The distance travelled by the vortex rings depends upon the duration of the flicks – the shorter the flick, the farther they go. However, at a critical point the air flow becomes turbulent and the vortex breaks down. The point at which this occurs depends upon the Reynolds number (Re) which is related to the linear dimension of the wing and to its velocity. It was empirically determined with the model that turbulent flow sets in between Re 1333 and Re 1600, depending upon wing length. The actual Re number achieved by signalling crickets was about 800 and it was suggested that there was no adaptive advantage in exceeding this figure due to the increasing uncertainty of forming good vortex rings. What is interesting is that the theoretical figure at which turbulence is expected to occur is only Re 600, much lower than that actually found. It is possible that the veining on the wings and the fringe of hairs on the edges of the wings assist laminar flow. This would offset the tendency for turbulence to set in, thus maximising the distance over which communication will be possible.

Vibration of the body and its appendages

Sound production associated with vibration or tremulation of the body, or some part of it, is widespread throughout the Arthropoda, and while there are ample descriptions of the signals, no very detailed studies have been carried out on the mechanisms involved. This is possibly because in the majority of cases no specialized structures appear to be made use of and the mechanisms do not merit intensive research. Because of the nature of these signals most, if not all, are propagated through the substrate. In some cases the substrate may itself be vibrated sufficiently to radiate sound, but in any case this airborne component has not been shown to be the effective signal.

Vibratory signals may form a very effective means of communication, and this is well illustrated by chloropid flies of the genus *Lipara*. The adults are small; between three and seven millimetres depending upon the species. The larvae of these flies form galls on the reed, *Phragmites*. When females hatch from the galls they do not disperse, in contrast to the males which fly from reed to reed in search of a mate. Males have the problem of locating virgin females on a stem which may be up to three metres long with only a 0.5 to 2 per cent infestation rate. To achieve this the males produce trains of species-specific vibration pulses which are transmitted along the stems and which females can perceive at a range of two metres. Virgin females reply to males with vibrations of their own, and a duet between the two ensues until the male locates the female. This method allows males to sample a large number of *Phragmites* stems with the minimum of energy expenditure and forms a highly adaptive strategy. The carrier frequency of the signals does not appear to have been measured but is quoted as being 'a few hundred hertz', and is thus in the same range as the signals of auchenorrhynchous Homoptera (planthoppers) which also use plant stems for transmitting courtship signals (Mook and Bruggemann 1968).

The mechanism whereby these signals are produced is not known but during song production abdominal vibrations can be seen and these are clearly transmitted to the substrate via the legs. The method may be similar to that in the Hawaiian drosophilid, *D. sylvestris*, in which the abdominal muscles are involved in song production. In contrast to species such as *D. melanogaster*, amputation of the wings does not abolish song, but this can be accomplished by cutting some of the abdominal muscles or by otherwise immobilizing the abdomen (Hoy, Hoikkala and Kaneshiro 1988). It is not clear if *D. sylvestris* is unique within *Drosophila* in this respect, as the role of the wings and the thoracic muscles has only been demonstrated empirically in a few of the one hundred or so species whose songs have been recorded. Indeed some species keep their wings folded during song production and little movement can be discerned. In addition, the songs of dewinged flies can be recorded so long as a pressure-sensitive microphone is used and the flies are in physical contact with the diaphragm. The possibility of substrate-borne transmission of courtship song in some members of the genus is therefore a possibility. This highlights the point that our ability to record a biological signal in a particular medium does not necessarily ensure that the animal is itself using the same medium.

Abdominal vibration is also found in Lacewings (*Chrysopa* species) in which, as in *Lipara* and some *Drosophila* species, duetting between males and females occurs. Because lacewings are able to hear ultrasonic frequencies it was thought initially that the important component of the signals might be a high-frequency one. However, such signals are unsuited to substrate-borne communication, and it has been shown that the significant component of the signal is below 100 Hz. The animals appear to choose thin laminar structures on which to call rather than

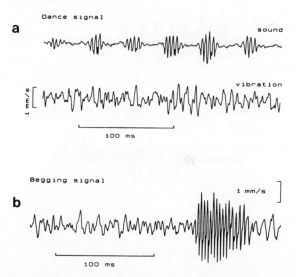

Figure 2.7. a, Waggle dance signals of a worker honey-bee. A comparison of the air- and substrate-borne signals shows that communication is via the former. Begging signals, on the other hand (b), are transmitted as vibrations through the comb. (From Michelsen, Kirchner and Lindauer 1986).

solid ones which would transmit such vibrations less effectively (Henry 1980).

Social insects such as honey bees and ants for a long time have been known to use vibrational signals for communication within the colony. In addition to the air-borne signals produced during the waggle dance, follower bees produce 'begging' signals which induce the dancing bees to regurgitate a sample of food. These begging signals are produced by thoracic vibrations and are transmitted to the comb by the legs. The signals, as revealed by laser vibrometry, consist of pulses up to 100 ms in duration and with the maximum energy around 320 Hz. At this frequency the amplitude of the vibration is 20 dB above the ambient noise level of the comb (Figure 2.7). By artificially vibrating the comb with a probe it could be demonstrated that the bees are maximally sensitive to signals in the range of 300 to 400 Hz when these are transmitted through the substrate (Michelsen, Krichner and Lindauer 1986).

The behavioural assay used in the measurement of threshold values to vibration was the freezing response of the bees. This is also elicited by vibrational sounds called 'tooting' and 'quacking' which are produced by queen bees. Again the wing muscles are involved and the queens press themselves against the surface of the comb so that the thoracic vibrations are transmitted directly to the substrate (Figure 2.8). These calls are frequency modulated and contain a number of strong harmonics in

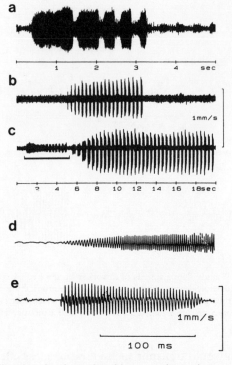

Figure 2.8. Vibration signals produced by queen honey-bees. a, Tooting; b,
Quacking; c, A quack induced by a toot (marked by the bar); d,e, Expanded
records of individual pulses of tooting and quacking to show the fine
structure. (From Michelsen, Kirchner, Anderson and Lindauer 1986).

addition to the fundamental of between 300 and 400 Hz (Michelsen,
Kirchner, Andersen and Lindauer 1986).

The honeycomb is a complex and inhomogeneous structure, and it is
therefore difficult to investigate its properties with regard to the transmission of vibrational signals. However, if the comb is artificially
vibrated it can be demonstrated that the attenuation of signals in the
frequency range of begging and the queen signals is particularly low.
This means that signal production is not only energetically efficient but
also that the intensities of the signals need to be carefully regulated in
order to limit the range over which they act.

Both web-building and errant spiders use vibrations of the body and
legs to transmit signals through the substrate or strands of the web.
Species of *Coelotes, Tegenaria, Amaurobius* all use webs in this way
during courtship, with both sexes participating in the exchange of
signals. These are pulse trains, mainly produced by drumming of the
pedipalps, and tone bursts resulting from abdominal tremulations. The
latter have a carrier frequency of between 8 and 150 Hz, depending upon

species and behavioural context. The method used to record these signals by means of a piezoelectric pick-up does not provide accurate information in the frequency domain. However, studies carried out in relation to prey capture by spiders using Laser Doppler Vibrometry, which does not load the web, provide useful information about their transmission properties. These demonstrate that signals are most efficiently transmitted by longitudinal waves which show little attenuation with increasing frequency and, indeed, show a degree of amplification between 3 and 10 kHz. It is at present unclear which components of the vibrations are used in discriminating between different sources of disturbance (Krafft 1978, Masters and Markl 1981).

The male Wandering Spider, *Cupiennius salei*, lives on bananas and other broad-leaved tropical plants. It produces trains of pulses, which have a frequency of 76 Hz, by vibrations of the body and legs. These signals are perceived by females at a range of up to one metre, and they respond with a burst of high-amplitude vibrations produced in the same manner as the male. At the point where the petioles diverge from the main stem, males have been observed to position themselves with their legs in contact with several petioles at once, enabling them to identify the one containing the female (Rovner and Barth 1981). For a small arthropod to transmit such signals satisfactorily requires good contact with the substrate. Some sparassid spiders have adhesive tarsal hairs which serve this function and it is probable that some such mechanism is widespread in spiders and insects.

In a similar fashion to *C. salei*, males of the tettigoniid, *Cophiphora rhinoceros* perform body tremulations consisting of bouts of between three and five pulses. These vibratory signals are perceived by females who may tremulate in answer. Males often locate themselves at the midrib of large-leaved plants which may possibly channel the vibrations. The male–female duet that ensues allows the male to locate the female. It is interesting that *C. rhiniceros* often alternates tremulation with stridulation which produces airborne sound. While the latter is more appropriate when the insects are on different plants, tremulation has the advantage that it will not be perceived by potential predators or competing males (Morris 1980). This strategy is probably common in other tettigoniids, as both *Neoconocephalus* and *Ephippiger* have been observed to tremulate as well as stridulate.

Body vibrations can also transmit signals on the surface of water, and this method is used by water striders (Gerridae). They produce the signals either by grasping some solid object with their forelegs and 'bobbing' the body, or while moving freely on the surface of the water. The vibrations are transmitted by the legs and result in a pattern of concentric waves. The attenuation of surface waves is considerable, some 10^{11} times higher than in water itself, and higher frequencies are preferentially attenuated. Water striders can detect these signals up to 10 cm and, at this range, most of the energy of the signals is concentrated below 50 Hz. This correlates well with the actual signals which are produced as these are between 17 and 29 Hz. Those arthropods which make use of surface waves for prey capture, such as the spider, *Dolomides triton*, and the back swimmers (Notonectidae), are capable of complex discrimination based upon different temporal and spectral properties of the signals. However, although gerrids are probably also capable of such discriminations it is not known if they make use of information other than the carrier frequency of the waves.

TYMBAL MECHANISMS

The songs of cicadas are among the most obtrusive of all arthropod sounds. In parts of the Mediterranean and in tropical areas they constitute the major source of biological sound, partly because of the habit of some cicadas to form choruses of many individuals and because the sounds themselves are among the loudest of insect songs.

The sounds are generated by a tymbal mechanism first described by Réaumur, in 1740. Tymbals are, however, found not only in the cicadas, but also in many of the smaller members of the Homoptera and Heteroptera, and in moths of the families Arctiidae (Tiger Moths), Ctenuchidae and Pyralidae. In addition, tymbal-like mechanisms, in which the sounds are produced by the buckling of an area of cuticle, have been described in cicadas and other Hemiptera, in addition to more usual tymbals, and in some nymphalid butterflies.

Homoptera: Cicadas and their relatives

The tymbals of cicadas lie on the dorso-lateral surfaces of the first abdominal segment. They consist of ridged areas of cuticle backed by an air space derived from the abdominal air sacs. A large muscle, the tymbal muscle, is inserted on the mesial edge of the tymbal from where a strengthening rib of thickened chitin runs to the posterior edge of the tymbal. The other end of the muscle attaches to a knob on the posterior of the metathorax. When the muscle contracts, it pulls on the rib which acts as a lever and causes a progressive buckling of the tymbal starting from the posterior edge. The stable resting position of the tymbal is convex, and when it buckles it clicks from the stable position to become concave; as the tymbal muscle releases tension the tymbal clicks outwards to resume a convex shape. One or both of the clicks may produce pulses of sound (Figure 2.9).

Detailed examination of the sounds of some species reveals that the sound pulses are double. Pringle (1954) originally interpreted this as being due to a quiet inward movement of the tymbal, immediately followed by a much louder outward recovery stroke. This seems

a

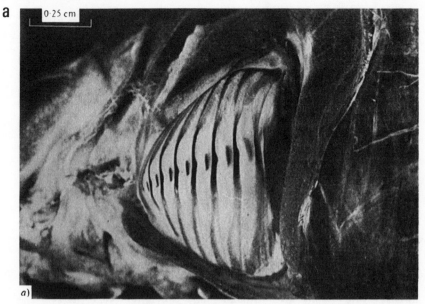

0·25 cm

a)

b

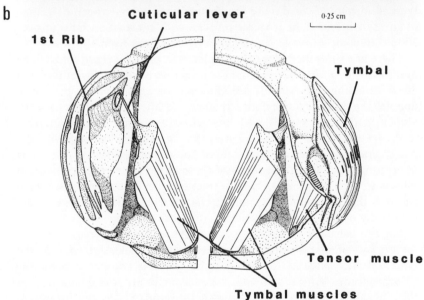

Cuticular lever

0·25 cm

1st Rib

Tymbal

Tensor muscle

Tymbal muscles

Figure 2.9. a,. The tymbal of the cicada *Cystosoma* showing the thickened ridges; b, An anterior view of a section through the sound-producing structures at two slightly different positions. The view on the left is through the second cuticular rib and shows the tymbal muscle with its lever attachment to the posterior edge of the tymbal. The more anterior view on the right shows the entire tymbal and its tensor muscle. (From Simmons and Young 1978).

unlikely, as most of the energy is applied in producing the initial click. It has been shown that in the Bladder Cicada, *Cystosoma saundersii*, the double pulse occurs because the tymbal buckles in two discrete stages, each producing a click, and the recovery is relatively quiet. It would seem likely that this is the general pattern and that Pringle was misled by using freshly dead material and artificially manipulating the tymbals to produce sound (Simmons and Young 1978).

There are three major factors which affect the nature of sounds produced by different species of cicada. These are the rate of contraction of the tymbal muscles, the detailed structure of the tymbals and the resonant frequency of the sound-producing apparatus. The repetition rate of the pulses is controlled primarily by the first of these, and may vary from as low as $40 s^{-1}$ in *C. saundersii* up to $500 s^{-1}$ in *Platypleura octoguttata*. In those species with very high rates of muscle contraction, the control of the muscles has been shown to be myogenic, as in the indirect flight muscles of the more advanced insects. Anatomical evidence suggests that the tymbal muscles may have evolved from the flight muscles. In most species that have been examined, however, muscle contractions are initiated neurogenically. Nevertheless, the tymbal muscles of cicadas are the 'fastest' that have yet been described of any synchronous, neurogenic muscle, with contraction rates of over $200 s^{-1}$ in some *Psaltoda* species (Young and Josephson 1983a).

There are a number of ways in which this primary pulse rate can be multiplied. In *Fidicula*, the rate of contraction of the tymbal muscles is $300 s^{-1}$ but the two tymbals are $180°$ out of phase, which effectively doubles the pulse repetition rate to $600 s^{-1}$ (Aidley 1969). This happens also in the *Psaltoda* species, and thus, while the muscle contraction rates in *P. argentata* and *P. claripennis* are $175 s^{-1}$ and $228 s^{-1}$ respectively, the pulse repetition rates of their calling songs are exactly double. It is interesting that the protest songs of these species have pulse repetition rates which are the same as the contraction rates of the muscles, which suggests that the animals are able to change from synchronous to alternate firing of the tymbal muscles.

The tymbals of many species possess ribs and, as the tymbal buckles, these may collapse sequentially, starting at the posterior edge of the tymbal. In *C. saundersii*, for example, this occurs in two stages, but in the 17-year cicadas (*Magicicada* species) twelve ribs are found, and nine of these buckle to produce nine sub-pulses of sound (Young and Josephson 1983b).

The carrier frequency of the songs depends upon the mechanical properties of the tymbals and the associated air sacs. In many species the individual sound pulses decay gradually, indicating the presence of a tuned resonant cavity which is being loaded. In the Bladder Cicada the abdomen is grossly modified to form a large air space. Theoretical

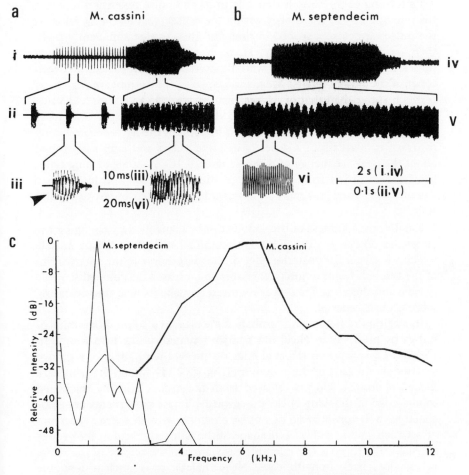

Figure 2.10. Oscillographs of the songs of *Magicicada casini* (a) and *M. septendecim* (b) at different sweep speeds to show the structure of the songs. The individual pulses of *M. cassini* song can be seen to be made up from nine sub-pulses (arrow) which result from progressive buckling of the tymbal. The resonance of the system in *M. septendecim*, by contrast, causes the impulses to be smoothed out; c, The frequency spectra of *M. septendecim* and *M. cassini*. The sharp tuning in the song of the former is clear. (From Young and Josephson 1983b).

calculations on the effects of this adaptation suggests a Q value of about 10 for the system, which would result in a consequent increase in sound pressure of 20 dB compared with a non-resonant system (Fletcher and Hill 1978). In the Bladder Cicada the resonant frequency is relatively low, 800 Hz, but in *Platypleura capitata*, which has much smaller air sacs, Pringle has calculated the resonant frequency to be between 4 and 5 kHz, a range which embraces the naturally occurring song frequencies.

Patterning of the sounds depends upon the pulse repetition rate and the rate of decay of the energy within the pulses, that is, the Q value. If the pulses are closely spaced in time and the Q value sufficiently high, then a continuous tone burst can be produced where the individual pulses are obliterated. This is well illustrated by a comparison between songs of the closely related *Magicicada* species. The tymbal of *M. cassini* is much stiffer than that of *M. septendecim* and has a very high Q value of 25 at the resonant frequency of 1.3 kHz. Its song therefore consists of an amplitude-modulated pure tone. The song of the *M. septendecim*, by contrast, is made up of a series of discrete pulses and is broadly tuned about the peak frequency of 6 kHz (Figure 2.10). A range of song types has been recorded from a variety of different cicada species, from a smooth tone burst, through an amplitude-modulated burst, to a pulse train.

Finally, the loudness of the sounds can be controlled by changing the curvature on the tymbal. This is accomplished by the tensor muscle which runs from the posterior edge of the mesothorax to the anterior rim of the tymbal. Contraction of this muscle increases the convexity of the tymbal and therefore the power required to make it click, thus releasing more acoustic energy.

In addition to possessing tymbals some cicadas also produce trains of pulses by means of a 'click' mechanism situated on the fore wings. In *Cicadetta sinnatipennis* the anal edge of the fore wing fits into a groove situated on the side of the scutellum (Figure 2.11). When the wings are flicked open, the wing is released from the groove and the thickened caudal edge of the wing clicks downwards. These movements are associated with the production of a pulse of sound with an energy spectrum of between 4 and 12 kHz. The shape of the wing edge and of the groove are such that there is considerable resistance to the release of the wing, yet it slots back in with ease. Nevertheless, it is unclear how the movement can be repeated at the rate of 32 s^{-1} found in *C. sinnatipennis*. This species is distributed in Asia Minor, but wing clicking has also been described in related species from Australasia, and so it may be a common phenomenon (Popov 1981).

The majority of bugs in the superfamily Cicacoidea, which includes the cicadas, along with several families of small plant bugs and hoppers, possess tymbals. The tymbals of these small bugs are probably homologous with those of the true cicadas and function in a similar fashion. The most detailed description of the morphology of the tymbals and of the associated musculature is still that of Ossiannilsson, published in 1949. However, in spite of the widespread occurrence of these insects, there has been rather little recent work on their bioacoustics. Such research that has been carried out has been more concerned with behavioural aspects of the songs (Claridge 1985).

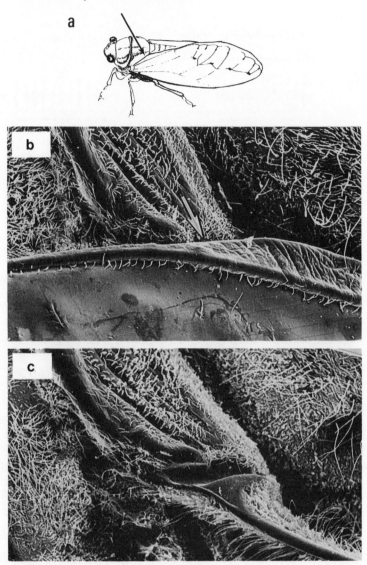

Figure 2.11. Click production in a cicada. The costal vein fits into a groove on the side of the pronotum. Its position is indicated by arrows in a,b, and the groove is seen in c, where the wing has been removed. A loud click is produced as the wing leaves the groove.

Tymbal muscles are present and one assumes, as in cicadas, that contraction of the muscles causes the tymbals to buckle inwards. One major difference is that the tymbals are not backed by large air sacs. If such sacs did exist then the carrier frequencies of the songs would have to be in the high kilohertz range to function as resonators. Unlike cicadas

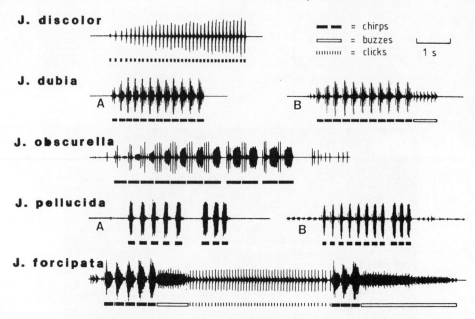

Figure 2.12. Oscillograms of the complex patterns of calling song, recorded from males of small plant bugs of the genus *Javesella. J. dubia* and *J. obscurella* each perform two distinct songs. (From de Vrijer 1986).

these bugs do not possess auditory tympani, and it is clear that the signals produced by the tymbals are transmitted through the substrate and not the air. For reasons already discussed this is a much more efficient means of long-distance communication than airborne sound for small insects. Where a spectral analysis has been carried out on the substrate vibrations the energy is seen to be concentrated below 1 kHz, which is appropriate for this type of signal. The calling songs of some *Javesella* species recorded using an acceleratometer are illustrated in Figure 2.12.

Heteroptera

Members of the second hemipteran suborder, the Heteroptera, also have tymbal-like structures. For example, in some of the pentatomids, cydnids and reduviids, the first two abdominal tergae are fused and modified for sound production. The cuticle contains a transverse furrow which, on contraction of some of the abdominal muscles, flattens. This movement is associated with vibrational pulses, although the precise mechanism of this 'tymbal' system does not appear to have been described. In the pentatomid, *Nezara viridula*, the vibrations take the form of downwardly frequency modulated pulses. Because of the dispersive nature of substratal propagation, the characteristics of the signal progressively change with distance from the source. In addition, the frequency components above 200 Hz become attenuated, while those below 150 Hz are amplified. The possibility therefore exists that the relative amplitude of different frequencies within the signals could be used by the receiver for distance discrimination, and there is some evidence that this may indeed occur (Cokl 1985).

Lepidoptera

The structure and function of the tymbals of arctiid moths have been described by Blest, Collett and Pye (1963) and show a remarkable degree of convergent evolution with those of the cicadas. The cuticle on the side of the metathorax (the katepisternum) of these moths is swollen outwards to form a blister of thin chitin which has, along its anterior edge, a number of striae. These vary in number from abut 10 to 60 depending upon species. The tymbals are distorted, apparently by contraction of the basalar muscle, and progressively buckle inwards starting at the dorsal edge. As in 17-year cicadas, the striae cause the buckling to occur in a series of discrete steps, each of which produces a pulse of sound. On relaxation of the muscle the process is reversed and a second series of pulses is produced. In the species studied, *Melese laodamia*, each paired tymbal movement or 'modulation cycle' had a periodicity of $40 \, \text{s}^{-1}$ while the individual pulses had an instantaneous rate of up to $1200 \, \text{s}^{-1}$. As each pulse produces a damped oscillation, the carrier frequency of the sounds is high, with much of the energy in the range of 45–90 kHz with a weak lower frequency component which can be heard by the human ear as a click. Sonographs of the modulation cycles show that the frequency is modulated in a stepwise fashion, dropping with each successive pulse and then increasing again during the recovery stroke (Figure 2.13).

A quite different tymbal mechanism is found in the Wax Moths (Galleriinae). The Greater Wax Moth, *Galleria mellonella*, bears a tymbal on the anterior part of each tegula, which is a small structure situated on the top of the fore wing, between the wing and the thorax. A blade of chitin, the tegular wing coupler, runs from the lower edge of the tegula and, during fanning of the wings, this is moved so as to cause the tymbal to click inwards. The outward recovery movement is due to the tension on the tymbal. Each wing depression thus produces two clicks which are of short duration, 100 µs, and very high frequency, 75 kHz (Spangler 1986).

The ultrasonic sounds of arctiid and ctenuchid moths may have an aposematic function with regard to bat predation, and the sounds produced by some nymphalid butterflies are similar in this respect. The Peacock Butterfly, *Vanessa io*, when performing its defensive display of wing opening, which reveals the eyespots on the dorsal surface of the

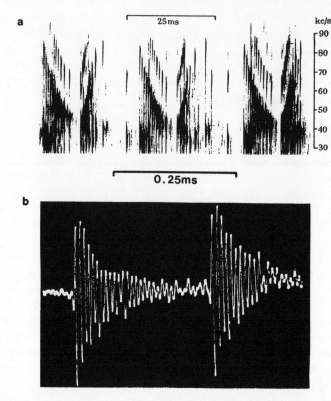

Figure 2.13. a, Sonogram of tymbal pulses of the moth *Melese*. The tymbals click out and then back in a stepwise fashion, producing a downwards and then upwards frequency-modulated pulse. b, Detail of the individual impulses. (From Blest, Collett and Pye 1963).

wings, produces very intense ultrasonic clicks which have a SPL of 100–110 dB at a range of 8–10 cm and a carrier frequency of 30–60 kHz. The clicks result from the buckling of a small bi-stable area of the fore wing near its base and between the costal and subcostal veins. This buckling may be initiated by the wing striking the tegula as it opens (Mohl and Miller 1976). Other Nymphalids possess similar mechanisms, including perhaps members of the neotropical genus *Hamadryas* (= *Ageronia*), which are well known for producing a series of loud clicks as they fly.

STRIDULATION

In arthropods, stridulation is the process whereby sound or vibration is produced by the friction of two body parts moving across one another. In human terms, running a fingernail along the teeth of a comb or playing a stringed instrument, such as a violin, are forms of stridulation.

Table 2.1. Stridulatory structures in arthropods

	Structures	*Reference*
Crustacea		
Palunirus	antenna:rostrum	Moulton 1957
Armadillo	propodite:pereiopod	Caruso and Costa 1976
Arachnida		
Therididae	prosoma:opisthosoma	
Salticidae	chelicerae:pedipalps	
Mygalomorpha	coxa:pedipalps	
Lycosidae	cymbium:pedipalps	von Starck 1985
Insecta		
Odonata	hind femur:abdomen	Asahina 1939
Orthoptera	femur:fore wing	
	femur:abdomen	Kevan 1954*
	tibia:fore wing	Dish 1961*
	tibia:abdomen	
	hind wing:fore wing	Riede 1987
Coleoptera	elytra:abdomen	Claridge 1974
	fore wing:elytra	Freitag and Lee 1972*
Mecoptera	juga:metanotum	Sanborne 1982
Leipdoptera	wing:leg	Hannemann 1956*
Siphonaptera	thorax:tibia	Smit 1981
Hymenoptera	abdomen:abdomen	Spangler and Manley 1978

The table contains examples which are not discussed in the main text and some reviews, marked *.

Some authors have used the term 'strigilation' in lieu of stridulation, but the former is a more limited term and appears to exclude those few cases where specialized stridulatory structures do not appear to be involved. Stridulation is a more general and widely used term.

This is an extremely widespread method of sound production in arthropods, and has been described in insects, arachnids and crustacea. Stridulation has evolved as a by-product of a variety of activities such as locomotion and feeding, and almost every possible combination of body parts which can be brought together and rubbed across one another is made use of by some group of arthropods. Table 2.1, in addition to the examples discussed in the text, gives some idea of the variety of structures involved and the widespread occurrence of stridulation in the Arthropoda.

The terminology used to describe the structures involved in stridulation is not standardized, and various authors use different terms for the same structures. Essentially two separate body parts must be involved.

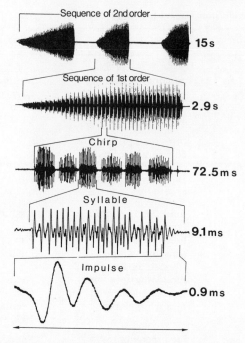

Figure 2.14. *Chorthippus biguttulus* stridulation to show the hierarchical structure of song and the terminology used to describe it. Each impulse is due to a single tooth strike. A syllable or pulse results from an entire leg movement, and a chirp (echeme or verse) is a repeated set of pulses. There is no general vocabulary for higher-order groupings. (From Elsner 1974).

One, which has been variously named the pars stridens, comb, strigil or file, possesses a series of corrugations, ribs or teeth, and the other, the plectrum, scraper or strigil, consists of a single peg, tooth or rib. Vibration is produced when the plectrum and file are scraped together. To be terminologically correct, the plectrum should move across the file but, depending upon the situation, one or the other or both of the structures may be moved to produce the vibration.

Before describing specific examples of stridulation there are some general considerations which need to be taken into account, and we require to establish a vocabulary for describing the parameters of the sounds that are produced. Each movement of the file and scraper results in a series of tooth strikes or impulses. The sound that is produced by this unitary movement is called either a pulse or a syllable. Often a series of such syllables occurs in rapid succession, and this series is termed a chirp, a verse or an echeme. Unfortunately, in the literature no single terminology has gained acceptance, and even individual workers are not always consistent in the terms that they use. Where the pulse repetition rate is sufficiently high, the human ear registers the chirps as single sounds and, if a large number of such pulses are produced sequentially, the stridulation is heard as a continuous, possibly amplitude modulated trill (Figure 2.14).

The form of the individual pulses will depend upon a number of variables. The tooth strike rate is affected by both the velocity of movement of the plectrum across the file and the spacing of the teeth on the file, while the duration of the pulse depends upon the velocity of movement and the effective length of the file. The carrier frequency and bandwidth of the signal are controlled in a complex manner by the tooth strike rate and the acoustic properties of the stridulatory apparatus and associated structures. An important consideration is the degree of resonance found in the acoustic system. If the file and plectrum do not form part of a resonant sound-radiating structure, then each tooth strike will produce a highly dampled impulse with a low Q value and containing energy over a wide range of frequencies. At the other extreme, if there is a resonator with a high Q value, the amplitude of the vibrations set up by each tooth strike will not become substantially attenuated before the onset of the next, and a 'purer' note will be produced. A special case of the latter is where the tooth strike rate matches the resonant frequency of the resonator. In this circumstance a pure sine wave will be produced without appreciable harmonics, except at the beginning and end of the sound pulse, when transients will occur.

Given these variables it is not surprising that a very large variety of sounds is produced by different species. Even within the repertoire of a single species different songs can exist in which the temporal patterning and spectral content can differ.

Femoral stridulation in grasshoppers

A good example of stridulation involving a basically non-resonant system is to be found in grasshoppers of the subfamily Gomphocerinae, in which sound-producing mechanisms have been studied, mainly by Elsner (1974) and his co-workers. These grasshoppers produce complex patterns of sound as part of their courtship displays. A row of pegs on the inner portion of the hind femur forms a file which is drawn across the raised medium radial vein of the tegmen (the sclerotized fore wings) which constitutes the plectrum. Each impulse or tooth strike produces a highly damped oscillation which is normally above the resonant frequency of the tegmina which radiate the sounds. Figure 2.14 shows how these impulses are built up in an hierarchical fashion to produce the complete song pattern in one of the species, *Chorthippus biguttulus*.

In order to relate the complex sounds produced by some of the species to the corresponding leg movements, it was necessary to monitor both simultaneously. Initially the leg movements were characterized by making use of the Hall effect. The underlying principle is that a magnet

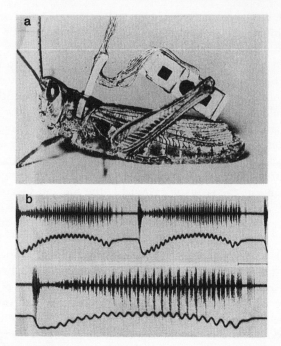

Figure 2.15. a, A Hall generator mounted on the femur of *Chorthippus mollis*, which provides a record of leg movement during stridulation.
b, Simultaneous recordings of sound (upper traces) and leg position (lower traces) during stridulation. (From Elsner 1970).

will induce a small transverse voltage in a current-carrying conductor, the Hall generator, which is situated close to it. A small magnet was therefore attached to the tip of the femur and the Hall generator to the thorax of the grasshopper. Movement of the femur during stridulation caused a fluctuating voltage which could be correlated with femoral position (Figure 2.15a). While this method has the advantages of simplicity and cheapness, the animal is somewhat hampered by the leads to and from the generator and, moreover, the movement of only one leg can be monitored. Subsequently, in order to overcome these drawbacks, an opto-electronic device was used in which a small piece of retroflective material which reflects a light beam back along its original path, was attached to the tip of the femur. The light so reflected was focused onto a photodetector whose output could be related to the position of the femur. This apparatus can be doubled up so that the position of both femora can be monitored at the same time. In addition, the animal is unencumbered by leads and magnets (Elsner 1974, von Helversen and Elsner 1977).

A record of simultaneous sound and femoral position during part of a courtship song in *Chorthippus mollis* is illustrated in Figure 2.15b. The

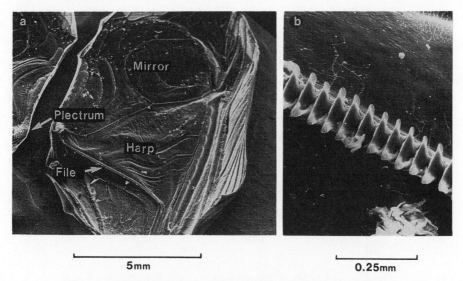

Figure 2.16. Stridulatory apparatus of the cricket *Gryllus bimaculatus*. a, A scanning electronmicrograph of the undersides of the elytra; b, High-power SEM of a portion of the file.

overall sequence commences with a few loud pulses, each of which results from a rapid, large-amplitude, up-and-down stroke of the leg. These are followed by a series of chirps consisting of a loud syllable, due to the downstroke of the femur, and then an amplitude-modulated vibratory phase where the alternate syllables are produced by a series of upward and downward movements. Initially, the upward strokes are longer, which results in the leg becoming progressively raised and then, as the downwards strokes lengthen, the leg is lowered again. Finally, a slow upward movement restores the leg to the starting position. Successive chirps become longer, that is, they contain more syllables, and the last few in the sequence omit the initial loud syllable of each chirp. To add to this complexity, the activity of the two legs is both asymmetrical and asynchronous. Thus, while one leg performs the pattern described above the other behaves in a similar way but omits the initial, loud syllables. In addition, the two legs act slightly out of phase, and this has the effect of 'blurring' the sound, a feature more pronounced in *C. biguttulus*, in which the structure of individual chirps becomes indistinct. This is an apparently inadaptive feature of the behaviour of most of the species studied, as the temporal parameters of the songs, which might be expected to be important in species recognition, are made less clear.

Elytral stridulation in gryllids
Stridulation using the elytra or tegmina is widespread in Orthoptera of the suborder Ensifera and acoustic regimes have been particularly well

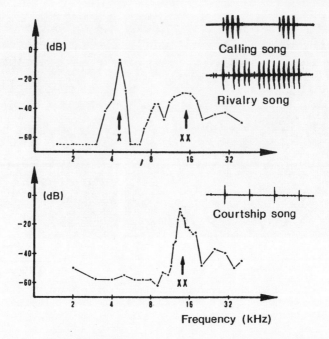

Figure 2.17. The three song types produced by the cricket *Gryllus campestris*. Frequency spectra show that in calling and aggression the sharpest peak is at around 4 kHz (x) with broader peaks at the second and third harmonics. For courtship song most of the energy is at the third harmonic of 16 kHz (xx). (From Nocke 1972).

studied in some of the field crickets (Gryllidae) and the bush-crickets and katydids (Tettigoniidae).

In gryllids the file consists of a row of chitinous teeth borne on the ventral side of elytral vein number 7 and the plectrum is formed from the inner edge of the opposing elytrum. The teeth on the file are angled like saw teeth so as to provide resistance when they strike against the plectrum. The elytra are usually relatively symmetrical, although the right elytrum is normally uppermost and sound is produced when the plectrum of the left elytrum is scraped against the file of the right one which lies above it during the closing stroke. If the positions of the elytra are artificially reversed, the animal can still stridulate, although the sounds produced are rather irregular. The opening stroke of the elytra is normally silent or almost so (Figure 2.16).

In the most studied species, the European Field Cricket, *Gryllus campestris*, three different song patterns are produced depending upon behavioural context. These are *calling*, *courtship* and *rivalry* or *aggression* (Figure 2.17). The songs differ both in respect of chirp repetition rate and the carrier frequency of the individual pulses. The calling and rivalry songs have a carrier frequency of between 4 kHz and 5 kHz while that of the courtship song is 16 kHz. The biophysics of sound production in this species, which we can use as a model for many other gryllids, has been studied by Nocke (1971, 1972).

The elytra in crickets are divided into discrete cells bounded by the veins. Nocke showed that the most important of these for radiating the 4–5 kHz component was a triangular area called the *harp*. The physical dimensions of the harp suggest a theoretical resonant frequency just below 5 kHz. In fact, by causing the harp to resonate sympathetically by using applied pure tones of different frequencies, and observing the maximum displacement, Nocke showed that the actual resonant frequency was 4 kHz. The relatively small deviation from the calculated value is probably due to the irregular shape of the harp, the calculations being based upon an idealized circular disc. Direct measurement of the sound emission from different regions of the elytra confirmed that most of the energy in the 4–5 kHz range is indeed radiated from the area of the harps. Ablation experiments reinforced this conclusion, as removal of both harps decreased the SPL by 46 dB. During stridulation both harps are set in motion and both elytra contribute to radiation of the sound.

A frequency power spectrum of the sound pulses shows that most of the energy is concentrated in a narrow band around the resonant frequency of the harp with smaller peaks at the second and third harmonics (8 and 16 kHz, Figure 2.17). It is clear therefore that we are dealing with a resonant system with a high Q value as is found in the songs of some of the cicadas. The calculated Q value for *G. campestris* calling song pulses based on their rate of decay is $Q = 19$. This high value is achieved because the tooth impact rate matches the resonant frequency of the system.

The description given above only relates to the calling and rivalry songs. The biophysics of courtship song, whose predominant carrier frequency is 16 kHz, is not understood. No region of the elytra with this resonant frequency has been identified, but the lower Q value of courtship song pulses might indicate that the elytra or portions of them are made to vibrate in a non-resonant mode. While this is acoustically inefficient, it may not be too important, as the songs only have to function at short range.

A major problem associated with sound production in 'open' systems such as are found in crickets, is that of acoustic short-circuiting. This is the tendency for particle movement round the edges of the sound-radiating surface during vibration to tend to cancel out the pressure waves. One way in which this problem can be overcome is by the use of a closed box, as in many loudspeaker cabinets, and this is seen in the cicadas where the tymbals are backed by an enclosed air space. Another method is to make use of a baffle, and this is achieved to some extent by crickets. During stridulation the elytra are raised and form an open box shape which extends some way around the harps on two sides, with the body forming a third. For a baffle to function adequately it is necessary for it to extend for a distance of at least 1/4 the wavelength of the emitted sound. A 4 kHz tone has a wavelength of about 8 cm, and so a truly effective baffle would need to be 2 cm. In fact the distance from the edge of the harp to the border of the elytra is less than 1 cm, and the baffle will therefore be only partially effective. One imagines that there is a trade-off between acoustic efficiency achieved by producing higher frequencies and the greater attenuation which this would result in.

Elytral stridulation in tettigoniids

Superficially a very similar mechanism to that described above in gryllids is found also in bush-crickets (Tettigoniidae). In most species the elytral structures used for stridulation are more specialized than in gryllids, as the two elytra are not symmetrical. The right elytrum bears, on its median edge, the plectrum, and also a structure called the mirror. Gryllids also possess this feature, which lies adjacent and posterior to the harp, but it has not, in them, been shown to be important in sound radiation. The mirror is a raised area of thin cuticle bounded partially or completely, depending upon species, by thickened veins collectively called 'the frame'. The left elytrum often has no specialized mirror region and bears the file which moves over and across the underlying right elytrum during stridulation

By ablating some of the teeth from the file and recording the sound pulses produced thereafter, Suga (1966) was able to relate the details of the stridulatory movements with the actual sounds produced in several different species. Removal of teeth caused either a local diminution in sound intensity or a silent period. In most species, as in the gryllids, only the closing stroke produced sound, although in one, *Conocephalus saltator*, a sound pulse was also associated with the opening stroke.

Suga was also able to demonstrate that the number of cycles of sound produced by single tooth strikes varied with species. In *Drepanoxiphus modestus* there appeared to be a 1:1 relationship between the two providing, as in Gryllidae, a high Q system, with each syllable consisting of a pure tone burst of 22–24 kHz. However, the more common system which he found in *C. saltator* and two *Phlugis* species was that each tooth strike resulted in a discrete damped pulse of sound with characteristically low Q. In these, as expected, the sonograms showed a wide band of frequencies. Ablation of large areas of the elytra had little effect on the frequency of the sounds, and it appeared that the region of the mirror was the important structure concerned with radiating sound.

It is impossible to gain a detailed understanding of the mechanism of sound production from the simple ablation experiments carried out by Suga on stridulating animals. Consequently, Bailey (1970) and Bailey

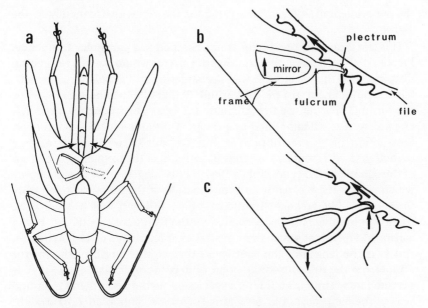

Figure 2.18. Stridulation in *Ruspolia* (= *Homococoryphus*) *nitidulus*. a, The posture during stridulation; b,c, Diagrams to illustrate the movement of the mirror during sound production (arrows). During the closing stroke the teeth on the file deflect the plectrum upwards and the mirror moves down, pivoting about a fulcrum. As a tooth disengages the plectrum, the mirror is restored to its equilibrium position due to the elasticity in the system.

and Broughton (1970) removed the elytra from the Tettigoniid, *Homorocoryphus* (= *Ruspolia*) *nitidulus*, and mounted them on a jig so that the file and scraper could be actuated mechanically in a controlled fashion. The sounds produced in this way closely resembled the natural song, and the method allowed the mechanics of the system to be observed and various input parameters to be changed.

The mechanism is best understood by referring to the diagrams shown in Figure 2.18. In *H. niditulus* the mirror frame is not complete, but U-shaped, being open laterally. The base of the U is connected to the plectrum by the vestigial file. As the file on the left elytrum is moved across the plectrum, the mirror frame is caused to vibrate, with the vestigial file acting as a lever about the fulcrum point where it attaches to the frame. Vibration of the mirror independently of the rest of the elytrum is facilitated by the thin cuticular membrane which partially surrounds the frame. Each tooth strike causes the plectrum to move upwards initially, thus levering the mirror frame down. As the plectrum loses contact with the tooth, the elasticity of the system restores it to the starting point and the cycle is repeated with each tooth strike. In *H. nitidulus* the tooth strike rate is $13\,000\,s^{-1}$, which corresponds well with the experimentally determined value for the resonant frequency of the elytra.

It is not clear how the mirror frame is tuned to a particular frequency. Pierce (1948) originally thought that the mirror would act as a resonating disc, because the species he was investigating, *Conocephalus spartinae*, has a rigid boundary to the mirror unlike that of *H. nitidulus*. However, the formula that he used to calculate the resonant frequency gave discrepant results when applied to a range of different species. Moreover, destruction of the membrane within the mirror resulted in a large reduction in SPL, but did not significantly affect the frequency spectrum of the sound (Morris and Pipher 1967). This suggests that it is the frame which acts as the resonator and the function of the mirror is to amplify the sounds. The resonant frequency of a cantilever, which is what the frame approximates to, is calculated by the formula $f_{res} = K/L^2$. K is a constant dependent upon the physical characteristics of the structure, and L is the length of the cantilever, that is, the length of the mirror frame from the fulcrum to the point of maximum displacement. If K is assumed to be similar in different species of bush-cricket then f_{res} plotted against the reciprocal of L^2 for a range of species should fall on a straight line. Sales and Pye (1974) figure such a graph for some 18 species, with the data derived from a number of different sources, and a fairly convincing linear fit is seen. This is not in itself a proof of the cantilever hypothesis which still requires experimental testing.

The majority of tettigoniids so far looked at produce sound in the ultrasonic range and, while some have side-bands within the sonic range, others cannot be heard by man. Many species with low Q systems produce song with a wide-frequency range and with one or more substantial harmonics. Sales and Pye (1974), for example, give ranges of 20–90 kHz for *Conocephalus dorsalis* and 32–120 kHz for *C. saltator*. While the mirror frame may be tuned to the major energy peak for each species, we have little idea how the side-bands are radiated.

Acoustic efficiency of elytral stridulation

The song patterns of stridulating insects have been determined during the course of evolution by the relative importance of several conflicting pressures. The efficiency of sound production, in terms of the amount of muscular effort converted into acoustic power, is related to the carrier frequency of the sounds. For high efficiency to be achieved, the wavelength of the emitted sound must be small in comparison with the dimensions of the radiating surface. The high frequencies are achieved in part by the impulse multiplication that occurs in stridulatory systems, whereby the unitary movement of a leg or elytrum bearing the file results in a large number of tooth strikes. In addition, some of the stridulatory muscles in, for example, the tettigoniids, are capable of very high rates

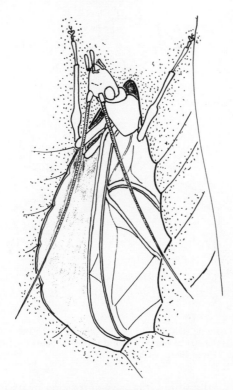

Figure 2.19. *Oecanthus* species stridulating from a hole it has chewed in a leaf. The leaf acts as an acoustic baffle to increase greatly the efficiency of sound production. (After Skaife 1979).

of contraction, second only to the cicada tymbal muscles (Josephson and Halverson 1971). However, opposing this tendency is the fact that high frequencies, especially those in the ultrasonic range, attenuate more than low frequencies, particularly in the presence of obstacles such as vegetation.

One method of improving acoustic efficiency is to make use of resonant systems as described above. Another is to increase the effective size of the baffle so as to make the acoustic pathway between the two faces of the radiating surface longer. This has been achieved in two different and most interesting ways by manipulation of the environment.

During stridulation some tree crickets (Oecanthinae) site themselves so that the leaves on which they sit act as partial baffles. They do so either by positioning their bodies at the edges of leaves, or by occupying holes chewed by other insects. One South African species, *Oecanthus burmeisteri*, actually chews a pear-shaped hole in a leaf, and stridulates with its elytra and fore body occluding the space (Figure 2.19). This has the

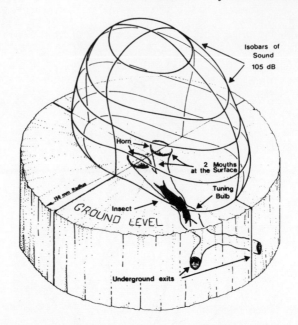

Figure 2.20.. The mole cricket, *Gryllotalpa vineae*, in the burrow from which it sings. The asymmetrical beam of sound probably helps the flying female to locate the male. (From Bennet-Clark 1975).

effect of increasing the sound intensity of its song by up to 10 dB. As this species has a rather low-frequency song of 2 kHz there will have been strong selection pressure to improve acoustic efficiency (Prozesky-Schulze, Prozesky, Anderson and van der Merwe 1975).

Another method is used by some burrowing crickets, in particular mole crickets, family Gryllotalpidae, which sing from within more or less complex burrows which they construct underground. *Gryllotalpa vineae*, for example, positions itself, head inwards, at the base of a double exponential horn whose branches join and then constrict just in front of the tegminae. The anterior part of the cricket's body occupies an enlarged space, the bulb (Figure 2.20). This effectively functions as an infinite baffle which means that the load upon the harps is entirely resistive. The exponential horns act as acoustic transformers which match the impedance of the air in the burrow to the sound-radiating surfaces of the tegminae. It is possible that the bulb is also important in increasing efficiency by tuning the horns. The effectiveness of this system is evidenced by a SPL in excess of 90 dB measured 1 metre from the burrow, while the songs can be heard at a range of 600 metres or so. The use of a doublet sound source, the two horn mouths, which are about one third of a wavelength apart, produces an asymmetrical sound field which is more efficient than a symmetrical one in attracting the females which normally fly in search of mates (Bennet-Clark 1970). A rather similar doublet sound source is produced by the nares of some rhinolophid bats to produce directional sonar cries.

Stridulation under water

Because of the impedance mismatch between air and water there is little transfer of acoustic energy between the two media, most being reflected at the interface. As a result we are usually unaware of the sources of biological sound in rivers and ponds. However, sound production by aquatic insects is widespread, and much of it results from stridulation. Stridulatory mechanisms are almost universal among the aquatic Homoptera and are found also in Coleoptera and the larvae of some Odonata and Trichoptera, while in the sea some decapod Crustacea are also known to stridulate (Aiken 1985, Field, Evans and MacMillan 1987).

Apart from descriptions of the sounds themselves, many of them subjective, and the identification of the structures that may be involved, rather little work has been carried out on underwater sounds. This is partially because many workers are unaware of their existence, and also because the apparatus for recording the sounds is not readily available. There are also technical difficulties in recording underwater sound in the laboratory, as the walls of an aquarium reflect the sound waves and the complex echoes can make interpretation of the signals difficult. In order to avoid this problem the investigation described below by Theiss (1982) was carried out with the corixids being restrained in a mesh net in a large body of water.

Corixids stridulate by means of a file on the femora of the front leg, which is rubbed across a plectrum situated at the side of the head. These movements generate pulse trains at up to $200\,s^{-1}$ whose carrier frequency is in the range of 1.5 to 2.8 kHz.

When these bugs dive, they trap a bubble of respiratory air which is located between the elytra and body on the dorsal side and also over the ventral surface. Theiss formed the hypothesis that the air bubble acted as a resonator, with the bubble pulsating during stridulation, due to vibrations of the head. It is possible to calculate the resonant frequency of a spherical gas bubble under water which is, in a simplified form, f_{res} (in kHz) $= 3.3/r$ where r is the radius of the bubble in mm. Fortunately, within quite wide limits, the shape of the bubble has little effect on its resonant frequency, and thus the effective radius of a non-spherical bubble can be calculated from its volume. Theiss was able to measure the volume of the air bubble in *Corixa*, and calculate the resonant frequency, which fitted well with the range of carrier frequencies actually determined from Fourier analyses of the sound pulses.

During underwater respiration the air supply becomes progressively depleted and, if the above hypothesis is correct, the resonant frequency should increase with time spent under water as the bubble becomes smaller. Theiss showed that there was indeed a linear relationship between time under water and carrier frequency. When a bug replenished its supply at the surface there followed an immediate decrease in the resonant frequency of the air bubble.

The relationship appears to hold across species and a linear plot was obtained when the main carrier frequency in the songs of different aquatic Hemiptera was plotted against body length. The latter is positively correlated with the size of air bubble carried. Small species such as *Micronecta*, with body lengths of around 2 mm, produce sound pulses in the 11 to 12 kHz range while, as noted above, *Corixa* species with 1 to 2 kHz songs are between 11 and 13 mm in length. It may be, therefore, that the resonant air bubble mechanism of radiating sound underwater may be widespread in Hemiptera and, possibly, in other aquatic insects.

Stridulation and substrate-borne vibration

In all of the above cases, sound produced by stridulation is transmitted through the fluid, either air or water. However, in many instances, particularly in smaller arthropods, the vibrations are transmitted through the substrate. Thus, many of the Hemiptera discussed above in the context of tymbal mechanisms also stridulate, and both methods may be used alternately within the same bout of song. The significance of this type of behaviour is unclear, but the stridulatory signals contain significant amounts of energy at higher frequencies than do the tymbal signals. However, we do not know what are the salient features of the songs used in determining the position, range and identity of the signaller. At present we can only speculate on the function of this apparently complex system of communication.

Vibrations transmitted through the substrate do not require complex sound radiators, such as the elytral modifications of tettigoniids and acridids. One general requirement, however, is for efficient coupling between the substrate and the stridulating animal. The tarsi and pretarsi of many insects and arachnids are pre-adapted for this function, due to their requirement for walking on leaves and other smooth objects. One specialized structure has been described in wolf spiders (*Lycosa* and *Schizocosa* species) which possess a stridulating organ in the tibio-tarsal joint of the palps. The palpal tarsus bears a group of spines or macrosetae which contact the substratum and probably have the function of transferring the vibration (Rovner 1975).

In the absence of specific sound radiators, any area of cuticle which vibrates will act as a sound source. The only study which has been carried out to investigate the radiation of sound from the general body surface is in the leaf-cutting ant, *Atta sexdens* (Masters, Tautz, Fletcher and Markl 1983). Many species of ant are known to stridulate by means of a file situated on the dorsal surface of the first abdominal segment, which is scraped across a plectrum on the posterior edge of the metathorax or postpetiole. While this produces audible sound, the behaviourally relevant channel for transmission is via the substrate, either along the surface or through it. Workers which become buried communicate by this means with other ants in order to recruit rescuers.

Laser Doppler vibratometry (LDV) was used to examine the modes of vibration of the gaster during stridulation, and two major vibrational components were identified. A 1 kHz component was related to the tooth strike rate and this was associated with an additional high-frequency, 10 kHz component. Microphone recordings showed that the wide-band sound pulses were produced with major peaks at these two frequencies. There are at least four different vibrational modes which could account for these features of the signals. The LDV study identified an up-and-down (dipole) vibration of the gaster to be the source of the 1 kHz component and a symmetrical (monopole) oscillation to account for the high-frequency one. Other possible vibrational modes, quadripole and surface-bending waves, for example, did not appear to be involved.

As LDV uses light, it was possible to repeat the measurements under water, so as to use a more dense medium, and mimic the situation of a buried ant. As was expected, the increased mass loading shifted the frequency peaks downwards, and the high-frequency component dropped to 7 kHz. It was calculated that this would drop further to around 3 kHz if the ant was underground. Masters *et al.* (1983) point out that underground there would not be the same impedance mismatch between the fluid-filled gaster and the surrounding earth or sand as there is in air, and thus energy transfer would be much more efficient.

These observations and calculations refer only to the way in which the gaster behaves during stridulation, and not to the actual transmission of vibrational waves through the substrate. However, soil tends to filter out the higher frequencies, and both of the major components of the stridulation would be expected to be transmitted favourably. In addition, leaf-cutting ants are known to be particularly sensitive to a range of frequencies between 50 Hz and 4 kHz which encompasses the actual range of signals produced.

3. Receptor Mechanisms

All sensory receptors for the perception of sound and vibration are, by definition, mechanoreceptors. Throughout the arthropods there is a very large number of different types of mechanoreceptors, but only a few have been implicated in communication. As demonstrated in the previous chapter, there exists a diverse range of mechanisms for the production of sound and vibration, and many of these have evolved independently of one another. This is mirrored by the variety of receptor types which have evolved in parallel with these mechanisms. It is convenient to categorize the receptors on the basis of the acoustic parameters that they are sensitive to.

HEARING

There are three classes of receptor mechanism involved in the reception of acoustic stimuli. These are, particle movement, pressure and pressure gradient receptors.

Receptors which are responsive to particle movement or displacement are often structures such as small hairs which are unaffected by pressure change (Figure 3.1a). As the speed of particle movement attenuates much more rapidly than the associated pressure changes, such receptors primarily function in the near field and their range is therefore extremely limited. On the other hand, because of the vectorial nature of particle movement, these receptors are inherently directional in contrast to pressure receptors. It is worth noting that the potential value of particle displacement for communication is much greater in water than in air. As the wavelength of a given frequency of sound is roughly five times longer in water, so also is the extent of the near field. Unfortunately, very little is known about particle movement receptors in aquatic arthropods. Fish, however, have been shown to use this parameter for the perception of low-frequency sound.

Pure pressure receptors consist of an enclosed space incorporating a moveable tympanum or piston which stimulates the sensory receptor cells (Figure 3.1b). Because pressure is scalar, most ears which respond to pressure changes are paired so as to provide directional information. The ears of many vertebrates and of some insects are of this type. It should be noted that the walls of the enclosure should be non-compliant, as are the tympanic bullae of mammals. In insects, however, the

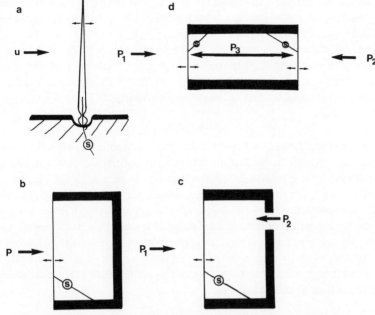

Figure 3.1. Different acoustic receptors in arthropods. a, A sensory hair
which responds to particle displacement; b, Pure pressure receptor; c,
Pressure gradient receptor. Displacement of the tympanum will depend upon
the difference in pressure at its two faces. This in turn will depend upon
several factors such as phase difference, length of the sound paths and the
physical properties of the leakage path; d, A pressure gradient receptor
where the tympani are coupled by an air space. Movements of the tympani
are critically dependent upon wavelength and direction of the sound.
Pressure waves can interact either destructively, and cancel one another out,
or constructively, so as to amplify the difference between the tympani.
S = Sense cells; u = particle displacement; P = pressure.

tympanic cavity may impinge upon air sacs or other deformable structures and this will affect the process of transduction.

Pressure gradient receivers are those in which the enclosure is to a
greater or lesser extent open, so that a sound pressure wave can act on
both sides of the tympanum (Figure 3.1c). The behaviour of such a
system is difficult to predict, as the deflection of the tympanum results
from the sum of the pressures acting upon it. This in turn depends on
the length of the sound channel between the two sides, the acoustic
impedance of the channel and the wavelength of the sound. It is evident
that, depending on direction and wavelength, the pressure may act
constructively or destructively at the two faces of the tympanum, thus
effectively amplifying or attenuating sound intensity.

In addition, if the sound path between the two faces is narrow, like an
insect trachea, the acoustic impedance will increase with increasing

frequency of a signal. This means that at low frequencies the tympanum will function as a pressure gradient receiver, but at higher frequencies, as a pure pressure receiver.

A special case of a pressure gradient receiver is where the enclosure is not open to the outside, but the two tympani form part of the same cavity. Each tympanum will be influenced, not only by sound reaching it directly, but also via the contralateral tympanum (Figure 3.1d).

VIBRATION

There are several means whereby animals are potentially able to perceive vibrations through the substrate. Movements of the substratum will be transmitted to the bodies of arthropods in contact with it. They may stimulate the proprioceptive organs which are found throughout the body, and particularly at the joints of the appendages. Vibrations will also cause distortion of the cuticle, and it is known that both insects and arachnids have receptors which respond to such stimuli. Man-made vibrometers work by monitoring the inertia of a mass when it is vibrated, and some body parts of arthropods such as the statocysts of crustacea could function in the same way.

RECEPTOR TYPES

Hair receptors

Virtually all arthropods possess surface mechanoreceptors in the form of fine cuticular hairs or threads. In the arachnids, these are the trichobothria; in insects, trichoid sensillae. Their basic structure consists of a chitinous hair from $100\,\mu$m to a few millimetres in length and a basal diameter of up to $10\,\mu$m. The hairs articulate with the surface by a socket or thin area of chitin which allows them to be deflected easily. The articulation may be such that the direction of movement of the hair is restricted, or they can act omnidirectionally.

The base of the hair is attached to the dendrites of sensory neurons which are stimulated by movement. While there are many such mechanoreceptive hairs which are not acoustic receptors, any that respond phasically are potentially capable of picking up airborne sound. In insects, each trichoid sensillum has a basic structure consisting of two cells, the tricogen and tormogen cells, which give rise to the hair and the socket, one or more bipolar sensory neurons and a neurilemma cell. The dendrites of the bipoplar neurons usually terminate in a non-neural structure, the scolopale or scolopid body which is involved in the transduction of the mechanical stimulus (Figure 3.2). Similar and possibly homologous structures are present in the Cructacea and Arachnida.

The properties of crustacean mechanoreceptive hairs have been examined, particularly those on the telson and the head appendages, and it is clear that they respond to water movement (Hawkins and Myrberg

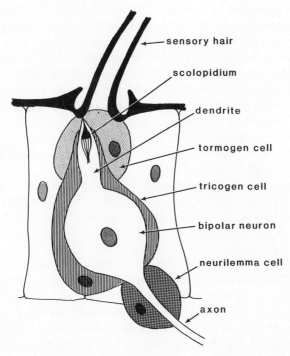

Figure 3.2. Diagram to show the structure of a generalized arthropod sensory hair.

1983). However, the sensitivity of the hairs appears to be too low for them to function as auditory receptors except, perhaps, at very close range. Nevertheless, crustaceans are known to respond behaviourally to underwater sound and it is possible that hairs, as yet uninvestigated, are responsible. The statocysts, which are primarily organs of equilibrium, could also be involved in sound reception.

The trichobothria of spiders are distributed on the walking legs and pedipalps and are often arranged in groups with a range of lengths. The optimal frequency to which these hairs could respond is related to length, and so the possibility exists for some form of frequency discrimination, although this has not been demonstrated so far. While there is evidence that trichobothria are used in the perception of particle movement, the extent to which such stimuli are involved in intraspecific communication is less clear. The majority of spider sounds also have a substrate-borne component to which spiders are known to be particularly sensitive (Barth 1982).

The biomechanics of hair receptors has been investigated only in detail for the thoracic hairs of the larva of the cabbage moth, *Barathia* (= *Mamestra*) *brassicae* (Tautz 1977), while a theoretical treatment of the topic is provided by Fletcher (1978). The results of their studies provide a general model for the behaviour of hair-like receptors.

The hairs of the caterpillar are approximately 500 μm long and have a resonant frequency of 150 Hz. Due to the high viscosity of the medium relative to the small dimensions of the hairs, their movement is highly damped and their Q value approaches unity. They are therefore broadly tuned and follow particle vibrations between about 100 and 400 Hz. In general, the resonant frequency would be expected to vary inversely with the square of their length. Their high-frequency response is limited by the necessity of the hair to emerge above the air boundary layer at the surface of the cuticle.

The range of frequencies over which filiform hairs can respond fits well with their known functions. In moth caterpillars they mediate defensive responses to the wing beats of potential wasp predators and parasitoids, while the trichobothria alert spiders to potential prey in the form of flying insects.

Hair receptors have been implicated in intraspecific communication in the cricket, *Phaeophilacris spectrum*, where the low-frequency wing flicks are picked up by the filiform cercal hairs (Kamper and Dambach 1979). In the Field Cricket, *Gryllus campestris*, and probably in other cricket species, the cercal hairs respond to the individual syllables within chirps which are repeated at $30\,\mathrm{s}^{-1}$. They do not, however, respond to the 4.5 kHz carrier frequency of the song. The cercal hairs are arranged in fields which have preferred stimulus orientations. Different groups of hairs respond to longitudinal and transverse deflection and thus provide directional information (Palka and Olberg 1977).

Johnston's organ

Another receptor which responds to particle movement is Johnston's organ, located on the second segment or pedicel of the antenna of many insects. It is stimulated by movement of the third segment which is often modified as a finely branched structure, the flagellum (Figure 3.3). The receptor itself is composed of up to several hundred chordotonal sensillae, each comprising a bipolar neuron with its associated scolopoid sheath and a cap cell which is stimulated by movements of the flagellum.

Johnston's organ has evolved primarily as a mechanoreceptor responsive to the tonic and phasic air movements that occur during active flight. As such it is pre-adapted for the reception of near field acoustic signals. Thus, for example, it is the antenna of *Drosophila* which is the receptor for courtship song (Ewing 1978) and the same is true for *Anopheles* mosquitoes (Tischner 1953). The range of frequencies over which the antennae of these diptera respond is similar, between 150 and 500 Hz. The resonant frequency of the antenna of *Anopheles subpictus* is 380 Hz,

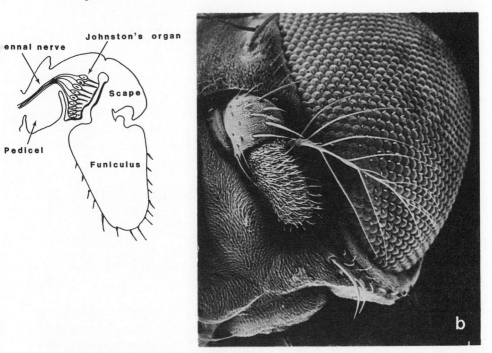

Figure 3.3. The antenna of *Drosophila*. a, Head, showing the position of
Johnston's organ: b, Section of Johnston's organ: The flagellum acts as a sail
to rotate the funiculus, thus stimulating the scolopidia of Johnston's organ.
(a, after Hertwick 1931).

which corresponds to the flight tone of the female and facilitates the
male's ability to locate her.

The use of vibratory signals produced by wing movement is common,
particularly in the Diptera, and Johnston's organ is likely to be used
widely as a near field acoustic receptor.

Tympanal organs

Moths. The simplest tympanal organs that have been investigated are
those that are found on the metathoracic segments of noctuid moths.
The primary function of these organs has been unequivocally shown by
Roeder (1966) and his co-workers to be the reception of the ultrasonic
cries of hunting bats. Similar organs of hearing are found in other
families of moths, including the Notodontidae and Arctidae. The
Geometridae and Pyralidae also possess tympani, but these are located
on the abdomen (Haskell and Belton 1956, Belton 1962).

The majority of these moth ears are employed as anti-predator devices
and almost all of the research has concentrated on this aspect of their
function. However, some also have a role in intraspecific communica-

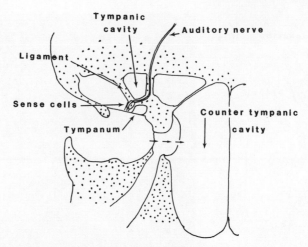

Figure 3.4. Dorsal view of a horizontal section through the tympanic organs and associated structures of a noctuid moth. The unshaded areas are all air sacs. (From Dethier 1963, after Roeder and Treat).

tion. Several species of moth have been shown to produce ultrasonic calling songs. The noctuid, *Heleothis zea*, does so by wing clapping while *Thecophora fovea* stridulates, and the Wax Moth, *Achroia grisella*, uses a tymbal mechanism (Agee 1971, Surlykke and Gogala 1986, Spangler, Greenfield and Takessian 1984). In arctiid and ctenuchid moths sound production may be used for attracting mates as well as warning off predators.

Although they have evolved independently on many occasions, tympanal organs share several common features. They are, almost universally, paired structures consisting of tympani backed by one or more air spaces derived from the tracheal system. Movements of the tympani are transduced by chordotonal sensillae whose scolopoid bodies terminate in cap cells (Figure 3.4).

In noctuid moths the tympani are separated by tracheal air sacs, and the possibility exists that movements of one tympanum could influence the other through internally transmitted pressure changes. However, this is unlikely to be a significant factor at the high frequencies at which these ears operate. While the sensory physiology of the noctuid tympanic organ has been studied intensively, little consideration has been paid to its biophysics, and it is not clear how the morphology of the tympani and associated structures is related to the frequency range.

In moths, the number of scolopidia associated with the tympanal organs is small; the notodontids have only one per side, the noctuids, two, with a third which is not directly concerned with the reception of sound, and the geometrids with four scolopidia on each tympanum. The small number of receptor cells reflects the fact that the ears are detectors of specific stimuli, i.e., the echolocating cries of bats, and are not required to make frequency discriminations. By recording the activity of the auditory nerves of noctuid moths, Roeder and Treat (1957) showed that they responded only to transients and that they were unable to discriminate pitch. The auditory system responded to a wide range of frequencies from 3 to 150 kHz with maximum sensitivity between 50 and 70 kHz. Sound intensity differences at the two tympani due to sound shadowing by the body and wings of the moths allowed them to evaluate the direction of the sound source (Roeder and Wallman 1966).

It would appear that the primary function of these hearing organs is the detection of bats, and that their use in intraspecific communication has evolved secondarily. In *T. fovea* the tympanal organs are maximally sensitive between 25 and 35 kHz, while the stridulatory signals have a carrier frequency of 25 to 30 kHz. These signals have evolved to conform with the existing properties of the receptor system.

Lacewings and mantids. Other, simple tympanal organs which appear to be 'bat-detectors' have been described for lacewings (Neuroptera) and praying mantids (Dictyoptera). In the former, they are situated at the bases of the fore wing radial veins. These are expanded dorsally to form an air space while the thin ventral cuticle forms the tympani, to each of which are attached two small groups of scolopidia. The frequency response of this organ is similar to that of the noctuids with a bandwidth of 13 to 120 kHz and a best frequency range from 30–60 kHz (Miller 1970, 1971).

Praying mantids possess a 'cyclopean' ear which consists of paired tympani in a cuticular recess between the metathoracic legs. The close apposition of the tympani suggests that they are not capable of directional hearing, and this is supported by auditory nerve recordings in which it was not possible to discern any directional component. Again, the evidence suggests that this is a bat detector, as the maximal response of the organ is between 25 and 45 kHz, and flying mantids have been shown to respond to bat cries. However, insufficient is known about the behaviour of mantids to rule out the possibility that they communicate acoustically (Yager and Hoy 1986).

Hempitera: cicadas. Within the Hemiptera the cicadas as well as the members of some other families have auditory tympani. In the former they consist of paired organs situated on the second abdominal segment below the tymbals (in the males). The chordotonal sensillae are enclosed in an auditory capsule coupled to the tympanum by an apodeme so that vibration of the tympanum stimulates the sensory neurons. The tympani are backed by an abdominal air sac which, in males, they share with the tymbals.

Recordings made from the auditory nerves of a number of different species show, not surprisingly, that the maximum sensitivity occurs at

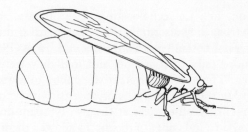

40 ms

Figure 3.5. Male of the bladder cicada, *Cystosoma saundersii*, showing the
greatly enlarged abdomen which is filled with air sacs which act as a
resonator for the calling song, an oscillogram of which is illustrated. (From
Simmons and Young 1978).

the carrier frequency of the conspecific song. In some cases the frequency tuning is sufficiently sharp for the ear to act as a peripheral filter
for the song, so that the carrier frequencies of the songs of heterospecific
cicada species are outside their ranges of hearing (Huber, Wohlers and
Moore 1980). Unlike those of moths, cicada tympanal organs are mainly
concerned with intraspecific communication.

A detailed study has been carried out on hearing in the Bladder
Cicada, *Cytosoma saundersii*. Although this species is rather atypical in
that it produces a low-frequency, *c.* 800 Hz song, it provides a model
which is relevant to other cicada species (Figure 3.5)(Young and Hill
1977). The morphology of cicada tympanal organs is relatively invari-
able between species, and that of the bladder cicada is exceptional only
because of the large air sac which, in males, almost fills the abdominal
cavity and provides the basis of its common name. Because the air sac is
involved in the production of sound, as well as its reception, the sac is
smaller in the silent females. For this reason hearing is rather different
in the two sexes.

Young and Hill (1977) recorded the auditory nerve responses and
showed that the auditory system was sharply tuned to the carrier frequency of the song at 800 Hz. In females, the response was highly
directional with a maximum difference in threshold sensitivity between

the two sides of 15 dB. The differential response was greatest for 800 Hz tones and decreased on either side of this figure to disappear below 600 Hz and above 1000 Hz. The SPL at a tympanum will depend upon the pressure impinging on it, both externally and internally, via the contralateral tympanum. The sound waves will interact either constructively or destructively, depending upon the phase of the sound. This in turn will depend upon the wavelength and direction of the sound and the length of the sound path via the contralateral tympanum. A simple model based on these factors and taking into account the body dimensions of the female cicada gives a cardioid, that is, 'heart shaped', directional response. The difference in sensitivity between the two tympani was found to be 15 dB for an 800 Hz tone, which agrees well with the physiological results.

Removing the posterior part of the abdomen exposes the tympanic cavity to the outside. In females, this results in a large decrease in auditory sensitivity across the entire frequency range, while in males the same procedure caused a marked decrease only around 800 Hz. The sharp tuning of the abdominal cavity in males is related to functions of sound production rather than reception. The ability of males to perceive the direction of a sound is actually decreased between 600 and 1000 Hz. This is because the entire cavity acts as a resonator in the form of a pulsating monopole at the carrier frequency during both the production and reception of song. In such a resonant mode, the sound pressure will always be in phase at the two tympani and, while this will enhance the sensitivity of the receptor system, it will mask the directional component of the sound.

The mechanisms responsible for the sharp tuning of the auditory responses in females and males are different because of the sexual dimorphism of the acoustic system. The tuning will depend upon the combined properties of the tympani and air sacs. It is noteworthy that in spite of these differences in the structure of the system, sound production in the male and its reception in females is tuned to the same frequency, that of the carrier frequency of the song.

Hemiptera: corixids. The other hemipteran tympanal organ that has been investigated is found in *Corixa punctata*. This is of particular interest as it is a corollary to the method of underwater sound propagation in these bugs which is described in Chapter 2.

Corixids have simple tympanal organs situated ventrally, at the bases of the fore wings. When the bug is submerged, the tympani do not come in contact with the water, but lie in a branch of the respiratory air bubble. It is likely that sound pulses transmitted through the water will cause pressure changes within the bubble and so stimulate the receptors indirectly.

Each of the tympanal organs has two scolopidial cells, of which only one is tuned to the carrier frequency of the conspecific songs of around 2 kHz. A bilateral asymmetry exists whereby one of the cells has a best frequency of 2.35 kHz and the other of 1.73 kHz. This is due to different resonant properties of the two tympanal membranes (Prager and Larsen 1981).

Prager and Streng (1982) determined, over time and at different temperatures, the resonant frequency of the air bubble comprising the physical gill of *Corixa*. The resonant frequency increased with time spent underwater from 1.7 to 2.7 kHz (at 12 °C) as the air in the bubble was used up. Temperature also affected the resonant frequency of the bubble with an average value of 2.4 kHz at 4 °C and 1.95 kHz at 16 °C.

As both sound production and reception are dependent upon the resonant properties of the respiratory air bubble, the changes that occur due to temperature variation will affect both the sender and receiver simultaneously and can therefore be neglected. However, signal matching could be a problem if the air bubbles have different volumes, and sharp tuning of the receptors could be disadvantageous. The different tuning curves of the two receptor cells could have evolved so as to ensure reception of the song over a wide range of frequencies.

Orthoptera. Tympanal organs are located in two different regions of the bodies of Orthoptera. In grasshoppers, locusts and other Caelifera the paired tympani are situated on either side of the first abdominal segment, while in the crickets and bush-crickets (Ensifera) they are found on the fore tibia. Considerable research has been carried out on these organs of hearing, both with regard to the biophysics of sound transduction and to neurobiology. Much technical ingenuity has been used in an attempt to relate the parameters of sound stimuli to the working of the orthopteran ear. That of the locust has probably been studied more intensively than any other organ of hearing in the arthropods. By comparison with the acoustic systems that have been described earlier in this chapter, hearing in grasshoppers and crickets is more complex. Some of the results that have been obtained are contentious, which probably reflects this degree of complexity.

Acrididae. The laterally placed abdominal tympani of locusts are separated internally by tracheal air sacs and a variable amount of body tissue. Direct measurement of the sound pressures at the internal surfaces of the tympani have shown that at low frequencies, i.e., below 8 kHz, sound is transmitted through the body from the contralateral tympanum. In these circumstances the ear will function as a mixed pressure and pressure gradient receiver, giving a directional, figure of eight pattern of hearing. As the locust is rotated through 180 °, the threshold for the auditory response, measured by the activity of the auditory nerve, varies between 4 and 14 dB, depending upon the individual animal. Heavier animals have poorer directional responses due to sound absorption by the body tissues between the ears.

At low frequencies, the wavelengths of the sounds are much larger than the dimensions of the body, and therefore the effects of diffraction will be negligible and will not contribute to the directional properties of the ear. At high frequencies, on the other hand, diffraction becomes important, and, at 15 kHz ($\lambda \simeq 2$ cm), an asymmetrical directional response is obtained. An intensity differential of 6 to 8 dB is found with maximum sound pressure occurring at 30 ° from the midline for the right tympanum (330 ° for the left) and a minimum at 180 °, that is, directly behind the locust. Above 10 kHz less than ten per cent of the sound energy is transmitted internally from the contralateral tympanum because of absorption due to the tissues and the increasing impedance of the air spaces. The body thus acts as a low-pass filter and, at high frequencies, the tympani function as pure pressure receivers which are inherently non-directional. Thus, the direction of high-frequency sound is perceived because of diffraction around the body (Michelsen 1971c, Miller 1977).

The ability of the locust to discriminate frequency was initially demonstrated by recording from the tympanal nerve and the connectives (Horridge 1961). Before then, it was generally considered that most insects were 'tone deaf'. There are two different methods whereby frequency coding can occur: the 'telephone' principle and the 'place' principle. The examples of frequency discrimination dealt with earlier in the chapter make use of the former principle. Thus, the arthropod hair receptors and the insect antennae vibrate at the same frequency as the particle movement. The sensory cells produce generator potentials which are in phase with the sound wave. Because of the limitation due to rates of firing in sensory nerves, the highest frequency that can be coded by this method in insects appears to be well below 1 kHz.

The place principle applies to vertebrate hearing where frequency is coded for along the length of the cochlea with the different hair cells of the basement membrane having different best frequencies. This is not a property inherent in the hair cells but is due to the mechanical properties of the system. The same principle applies to hearing in acridids. The locust tympanum is a complex and inhomogeneous structure consisting of a heavily sclerotized supporting rim, a large posterior region of thin cuticle and a smaller anterior area of thicker cuticle. The two are separated by sclerites which comprise, dorsally, the styliform body, ventrally, the folded body and, between the two, an elevated process. Posterior to this, and within the thin membrane, is the pyriform vesicle (Figure 3.6). The acoustic receptor, Müller's organ, which contains about 80 scolopidial cells, is attached to these sclerites.

The receptor cells are in four discrete groups (*a*, *b*, *c* and *d*) each

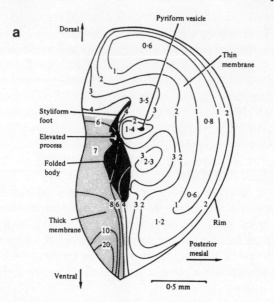

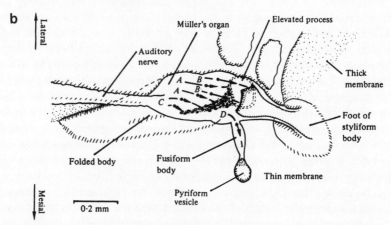

Figure 3.6. The tympanum of the locust. a, A map showing the position of the sclerites and the relative thickness of different regions of the tympanum; b, Detail to illustrate Müller's organ in relation to the sclerites and the position of the different groups of sensory cell bodies (A–D). (From Stephen and Bennet-Clark 1982).

connected to a different region of the sclerites, the *a* cells to the elevated process; *b* cells to the tip of the styliform body; *c* cells to the folded body; *d* cells to the pyriform vesicle (Stephen and Bennet-Clark 1982). Michelsen (1971a) showed, electrophysiologically, that the four groups of cells had different and characteristic best frequencies. By means of laser holography he mapped the modes of vibration of different areas of the tympanum, demonstrating that the thick and thin regions as well as the entire tympanum had different resonant frequencies. Michelsen suggested that this formed the basis for frequency discrimination of the ear with each of the four groups of sensory cells being attached to differently vibrating regions of the tympanum (Michelsen 1971b).

This original study failed to take into account the effect of Müller's organ, which has appreciable mass, and both dampens and modifies the tympanal vibrations while itself oscillating in a complex manner. A detailed examination of the behaviour of the entire structure made it possible to relate the distortions occurring at the receptor cells to specific frequencies (Stephen and Bennet-Clark 1982). The place principle for frequency discrimination is therefore confirmed, and there is no evidence that different sensory neurons vary in their sensitivities to frequency.

The range of best frequencies of the sensory cells is between 1.5 and 19 kHz, with some cells having more than one peak value. However, the composite input from the different groups of cells potentially resolves any possible ambiguity and allows complex frequency discrimination. The low Q value of the tympanic organ, along with the wide range of frequencies over which it is responsive, makes it well suited to perceive the broad-band stridulatory signals with their abrupt transients that are found in acridids.

Ensifera: Stenopelmatidae. The simplest ensiferan ear so far described is in the primitive wetas of New Zealand (family Stenopelmatidae). Their acoustic system bears a superficial resemblance to that of the gryllids and tettigoniids and therefore provides a good introduction to these more sophisticated types of hearing. In the Weta, *Hemideina crassidens*, there are two similar sized tympani (Figure 3.7a) on each of the fore tibia and these face to the front and back. They are backed by tracheal air spaces, and the chordotonal sensillae are contained in a sensory structure, the crista acustica, attached to the anterior tracheal membrane (Figure 3.7b). The leg tracheae are enlarged to form air sacs and they connect with the tracheae leading to the prothoracic spiracles. This effectively means that the pressure at the internal surfaces of the tympani will be the same (Field, Hill and Ball 1980).

The response of the acoustic organ, determined by the overall activity of the acoustic nerve, shows a best frequency of between 2 and 2.5 kHz with a steeper cut off above the optimum than below it. This correlates well with the spectrum of the stridulatory signals in this species.

One potentially important consideration in this system is the relative intensity of the sound arriving at the tympani directly from the source, and indirectly, via the thoracic spiracle. In *H. crassidens*, the latter appears to be largely irrelevant. Cutting the tracheal air sac midway along the femur and blocking it has no effect on the sensitivity of the tympanal response over the entire frequency range. Occluding the

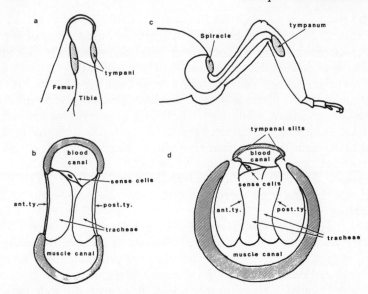

Figure 3.7. Comparison of the anatomy of Ensifern ears. a, Tibia to show the position of the two tympani; b, The Weta, *H. crassidens*, in which the tympani are similar in size; c, A bush-cricket: the tympani are within air spaces which open to the front through tympanal slits; d, diagram of the cross-section of a bush-cricket to show the connections between the thoracic spiracles and the tympani. (From Field *et al.* 1980, Larsen and Michelsen 1978).

tympani directly, on the other hand, causes a large decrease in sensitivity which is most marked at around the best frequency, where the response is reduced by 40 dB. Filling the auditory trachea in the leg by up to 60 per cent of its volume with water reduced the sensitivity of the ear, but only for frequencies below the best frequency. The auditory trachea therefore does not function as a resonator to determine the frequency response of the crista acustica, but acts as an amplifier for low-frequency sound.

The ear of the Weta can thus be considered as a single input pressure receiver and, as such, would be expected to be non-directional over its effective frequency range of between roughly 0.5 and 10 kHz. With the legs held away from the body, the effect of diffraction will be insignificant, particularly at the wavelengths at which the ear works best (at 2.5 kHz, $\lambda \simeq 13$ cm). Polar plots derived from the responses of the acoustic nerve confirmed this expectation with no variation in response with regard to the direction of incident sound for a wide range of frequencies (Field *et al.* 1980).

Ensifera: Tettigoniidae. The tettigoniid ear is similar in structure to that of the Weta. The tympani, instead of being on the outer surface of the tibia, are inside tympanic cavities which communicate with the outside by two slits angled latero-anteriorly and latero-posteriorly (Figure 3.7d). Unlike the Weta, the acoustic tracheae are not enlarged within the leg, but open out into large air sacs within the prothorax and connect with the prothoracic spiracles (Figure 3.7c). The shape of these tracheae varies between species but they are usually horn or funnel shaped. There are a number of related problems concerned with the way in which these acoustic organs operate; the function of the acoustic trachea in hearing, the possible roles of the tibial slits in directional hearing and the methods of frequency discrimination have all been the subject of research and some controversy.

Although several different hypotheses have been forwarded to account for the function of the acoustic trachea, all of the workers who have considered the problem agree that the tracheal input provides some acoustic gain at the tympanal organs. In many species, the tracheae have the form of an exponential horn with the diameter of the trachea progressively increasing from the tympanal organs in the tibia to the enlarged prothoracic spiracle. This suggests that the tracheae act as an acoustic transformer to match the impedance of the air with that of the tympani and thus increase the effective sound pressure. It is possible to calculate the properties of such a horn and thus predict the performance of the acoustic system. A series of partially contradictory experiments by Lewis (1974) on *Homorocoryphus nitidulus,* Nocke (1975) on *Acripeza reticulata* and Michelsen and Larsen (1978) on *Tettigonia cantans* have all failed to confirm predictions concerning the frequency response and gain of the ear. In addition, sharp resonances were found at particular frequencies when an exponential horn would be expected to be broadly tuned.

One difficulty with the interpretation of some of this earlier work is that auditory nerve responses were used to assay the gain in the auditory tracheae. This ignores the possibility that the receptor organs per se are tuned to a specific frequency range, different from that of the other physical components of the system, thus confounding the two factors. In an attempt to overcome this and other drawbacks, Hill and Oldfield (1981) inserted miniature microphone probes into the acoustic trachea within the femur and the thorax to measure the sound pressure levels directly. They looked at four different tettigoniid species and demonstrated a gain of up to 30 dB between the thoracic spiracle and the tympanum for a wide range of frequencies. In three of the species, the measured acoustic properties agreed well with those expected from exponential horns, while in the fourth, *Mygalopsis marki,* the gain was that expected from a conical horn. In this species the rate of taper of the trachea is more rapid and indeed resembles such a horn rather than an exponential one.

Hill and Oldfield also looked at the relative contributions of the spiracular versus tibial inputs for directional hearing. At frequencies below 8 kHz, the tibial inputs were involved in producing a directional response. Above this frequency, eliminating the tibial inputs to the tympani by covering them with a sealed tube did not affect directional sensitivity. It seems that at higher frequencies, differences in sound pressure at the thoracic spiracles are solely responsible for directional hearing. The difference in sensitivity of the two ears to a laterally occurring sound was between 10 and 20 dB. This difference is at least partially due to diffraction round the body, but also to the morphology of the spiracle and surrounding cuticle.

These results suggest that over much of the frequency range of hearing, inputs to the acoustic system through the tibial slits are relatively unimportant, a position held by some workers in the field. However, Stephen and Bailey (1982) have carried out some ingenious experiments to clarify the function of these structures in the tettigoniid, *Hemisaga* species. To investigate the way in which the slits functioned, they first had to null the effect of the thoracic input to the acoustic system. This they did by waxing a 1.5 mm internal diameter polypropylene tube to the prothoracic spiracle and bending it up above the knee so that its opening was in line with the tibial slits (Figure 3.8a). They then recorded the activity in the auditory nerve in response to 7.4 kHz tone bursts played at an angle of 90 ° to the body, and then progressively trimmed the end of the tube until the neural response reached background. This meant that when the preparation was rotated in the horizontal plane, the phase of the sound and its intensity entering the spiracle via the 'tuned' tube and the slits would be the same at all angles. Any variation in the neural response during rotation of the animal could thus be attributed solely to sound entering the tibial slits. A polar plot for this preparation is shown in Figure 3.8b, and shows clearly two lobes at 55 ° and 125 ° which correspond to the direction in which the tibial slits face. Increasing the frequency to 8.5 kHz is sufficient to abolish this directional effect (Figure 3.8c).

The tibial slits and the slit cavities with which they communicate probably act as Helmholz resonators which are sharply tuned to the carrier frequency of the song of *Hemisaga* which has a centre frequency of 7.4 kHz. At this frequency it is the slits alone that provide directional information to the ear; at higher frequencies the thoracic inputs have this function. Stephen and Bailey note that some species are able to close their prothoracic spiracles, and that the females do so during phonotaxis, thus eliminating the poorer directional response provided by the spiracular input.

The crista acustica of tettigoniids has most recently been described by Schumacher (1979) and comprises a linear array of 60 to 80 chordotonal sensillae whose scolopoid bodies and dendrites are inserted into a

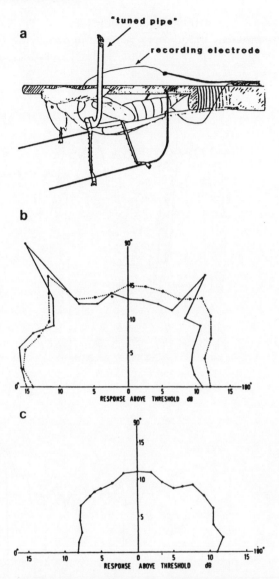

Figure 3.8. a, Tuned pipe preparation used to assess the function of the tibial slits in the tettigoniid *Hemisaga*. See text for an explanation; b, Polar response plot to an ipsilateral tone of 7.4 kHz in *Hemisaga*. Solid line from a normal animal. Dotted line, with the rear slit partially blocked; c, Increasing the frequency to 8.5 kHz abolishes the directional response. (From Stephen and Bailey 1982).

membrane bordering the vein to the leg (Figure 3.9a). Several workers have recorded from fibres within the auditory nerve and shown that the individual sensory units have different best frequency curves. As the cap cells and the dendrites of the chordotonal sensillae decrease in size from

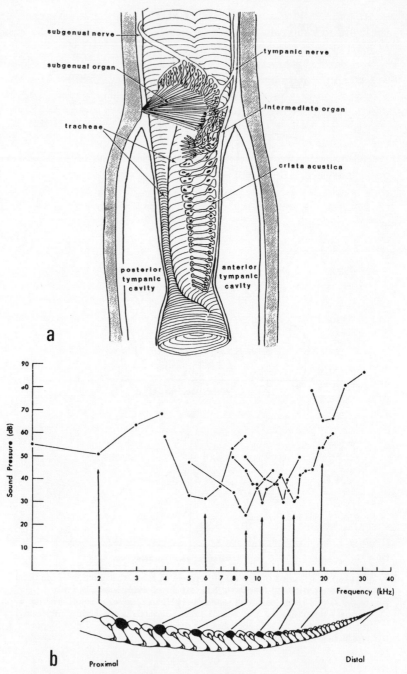

Figure 3.9a. Structure of the compound tympanic organ of a bush-cricket. (From Schwabe, 1906).

Figure 3.9b. Frequency threshold curves of 7 identified auditory sensillae in the *crista acustica* of *C. simplex*. (From Oldfield 1982).

the distal to the proximal end of the organ, the possibility arises that these are tonotopically arranged in a manner analogous to the basalar membrane of mammals.

By recording the neural responses of individual auditory sensillae and subsequently marking them with Lucifer yellow, Oldfield (1982) was able to confirm that this was so (Figure 3.9b). The best frequencies ranged from 2 kHz at the proximal end to 20 kHz at the distal end of the crista acustica, giving an overall frequency response of 1 to 35 kHz. It is noticeable that the absolute sensitivity of the cells below 6 kHz and above 18 kHz is less than for those between these two values. This is explained by the acoustic trachea itself being broadly tuned between 7 and 18 kHz, thus amplifying the response over this range. The basis of the sharp tuning of individual sensory cells has yet to be discovered, but does not appear to be due to the mechanical properties of the membrane in which they are suspended, as damage to the membrane does not significantly alter their frequency response.

Ensifera: Gryllidae. Arthropod hearing probably reaches the peak of complexity in the gryllids. In crickets, as in tettigoniids, there are both tibial and spiracular inputs to the acoustic system but, unlike the latter, the two sides are not independent of one another, and this adds a further dimension.

The tympani are paired and are at the surface of the tibia, as in wetas, and not recessed in tympanal cavities. One tympanum faces to the front and is smaller and thicker than the caudal facing one. It is not clear at present what part, if any, it plays in hearing. It is backed by about 100 μm of tissue which would effectively dampen its motion, and the experiments that have been carried out do not ascribe any clear function to it. The posterior tympanum is larger and thinner, backed by the acoustic trachea, and has been shown to vibrate in its basic mode, that is, like a piston, in response to frequencies between 1 and 30 kHz (Figure 3.10a)(Larsen and Michelsen 1978).

Several experiments, such as damaging or blocking the posterior tympanum, have shown that it is an essential link in the transduction of the acoustic stimulus. Vibration of the tympanum provides two possible types of effective stimuli, either pressure changes within the air space or the mechanical vibrations per se. By using two sound sources, whose amplitude and phase could be controlled independently, one driving the tympanum directly and the other indirectly through the trachea, Kleindienst, Wohlers and Larsen (1983) showed that it was movement and not pressure change which was involved. Thus, for example, when two tones of the same frequency, but 180 ° out of phase with each other were used, they effectively cancelled each other, and the tympanum did not vibrate. However, the pressure within the acoustic system would still oscillate, and yet no response was observed in the acoustic nerve.

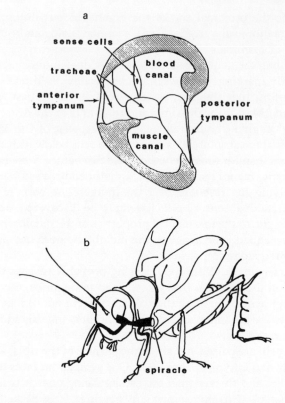

Figure 3.10. a, Diagram of the cross-section of a cricket leg to show the tympani and associated structures; b, The tracheal system involved in hearing, demonstrating that it is a four input system. (After Michel 1974).

The anatomy of the acoustic organ in crickets is similar to that of the tettigoniids with a crista acustica containing 25 to 30 sense cells. The individual sensory receptor cells in *G. bimaculatus* have been shown to be tonotopically arranged. Here also, the tuning of the receptors is not due to the mechanical properties of the tympanal membrane. Rather it appears to be an endogenous property of the sensillae, although the mechanical or electrical basis is unknown. The coupling between the complex acoustic events within the tracheae and the sensillae appears to be the tracheal membrane to which the crista acustica is attached.

The functioning of the cricket crista acustica has been shown to be similar to that of tettigoniids. Individual receptor cells have best frequencies between 3 and 11 kHz although it was possible to identify units in the acoustic nerve with responses up to 17 kHz. These may originate from sensillae lying distal to the 11 kHz units from which it was not possible to obtain recordings. Two groups of receptors with different thresholds which were selectively tuned to around 5 kHz were identified.

This concentration of units is likely to be related to the reception of the conspecific calling song (Oldfield, Kleindienst and Huber 1986).

The acoustic tracheae are not in the form of acoustic horns and do not appear to tune the system. Blocking or cutting the tracheae at various points has little detectable effect, either upon the sensitivity or the frequency discrimination as measured by the neural response (Hill and Boyan 1977).

The cricket ear is, at least potentially, a four input system, as the acoustic tracheae are separated in the prothorax by only a thin membrane which is acoustically transparent (Figure 3.10b). The behaviour of one tympanum will therefore be influenced by the external pressure and the internal pressure within the acoustic tracheae acting on it. The latter, in turn, will be affected by inputs to both spiracles and the contralateral tympanum. However, blocking the contralateral tympanum does not appear to affect either the threshold or the directionality of the response at the ipsilateral side (Boyd and Lewis (1983). The two spiracular inputs however, are important. Blocking the contralateral spiracle results in a reduction of directional sensitivity from 28 to 8 dB, while occlusion of the ipsilateral spiracle changes the directional response from a cardioid to a figure of eight one. Blocking both spiracles results in almost complete loss of directional sensitivity. In contrast to tettigoniids, the tibial input does not appear to provide the animal with directional information.

The ability of crickets to locate the direction of a sound source has been looked at in a number of different species. The finest discrimination is very sharply tuned to the carrier frequency of the conspecific calling song which is usually between 4 and 5 kHz. At such relatively low frequencies, body diffraction can offer little directional information. This is provided by the pressure gradient across the tympanum which varies because the phase and pressure within the acoustic tracheae depend upon the angle of incident sound.

Vibration receptors

Orthoptera. As early as 1941 Autrum demonstrated that the subgenual organs of insects were very sensitive to the acceleration component of substrate vibrations. These sense organs are to be found at the tibial-femoral joint of all six legs. They consist of up to about 25 chordotonal sensillae positioned under the cuticle at the front of the tibia with their dendrites within the haemolymph canal of the leg (Eibl 1978).

It has not yet proved possible to record directly from individually identified sensory cells, but one can record from single units at the point where the fibres leave the legs to enter the thoracic ganglia. On this basis, in the locust, the units can be classified into four distinct types (Kuhne 1982). Type I units respond to vibrations between 15 and 1000 Hz and

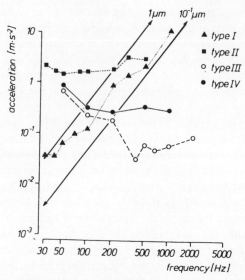

Figure 3.11. Threshold curves for the four types of vibration receptors in the fore leg of the locust. The straight lines indicate the accelerations resulting from displacements of 1 and 0.1 μ at different frequencies. (From Kuhne 1982).

the action potentials are phase locked up to a maximum of 200 Hz. Above this frequency the response becomes unpredictable (Figure 3.11). These receptor units respond to small displacements of between 0.1 and 1.0 μm but give a rather uniform response over a wide range of stimulus intensities. Individual units do not code adequately for intensity but, as different units within the leg have unique threshold sensitivities, their composite input effectively provides a large intensity range. Type I units are thought to be companiform sensillae which occur in groups on the cuticle of the tibia close to the sub-genual organ.

Type II, like type I receptors, respond to low-frequency vibrations but, in contrast to them, measure displacement and not acceleration (Figure 3.11). Their identity is not known but their properties suggest them to be chordotonal organs whose function is to register joint position, and they are proprioceptors rather than vibration receptors.

Type III units are the sensory cells of the sub-genual organ. Some of these are very sensitive to low levels of acceleration with thresholds below 0.01 ms^{-2}. They have an overall frequency range between 30 Hz and 8 kHz, maximum sensitivity being between 200 and 1000 Hz. Again, there is range fractionation among the units to provide coding for acceleration values from 0.01 to 1.0 ms^{-2}. Type IV units are also considered to be sub-genual in origin and are similar to the foregoing except that they have no clear frequency threshold. They also extend the upper range of acceleration intensity and, as they respond phasically, they can code for stimulus duration.

These four classes of receptor are also found in the tettigoniid, *Decticus verrucivorus*. However, in tettigoniids (and gryllids) the fore legs are different from those of the acridids in that they bear the tympanic organs. A complex sense organ contains the crista acustica, the sub-genual organ and the intermediate organ of unknown function (Figure 3.9a), and is therefore concerned with the reception of both air-borne and vibrational stimuli.

An additional feature found in *D. verrucivorus* was a third type of sub-genual receptor, type V, which was similar to type II except that the units were more sharply tuned to the higher frequency of 2 kHz (Figure 3.12). Both types III and V responded, not only to vibration, but to airborne sound as well. The thresholds for the latter were around 5 kHz and thus well above those of the vibrational responses. Finally, some of the type IV units responded to both vibration and to sound between 5 and 16 kHz (Kalmring, Lewis and Eichendorf 1978).

While all the details of this complex system are not yet entirely clear, and the identity of some of the sensory units which give rise to the responses is conjectural, we can see the ways in which a wide range of signals, acoustic and vibrational, from a few hertz up to 30 kHz is coded. Thus, the campaniform sensillae respond to vibrational signals below 200 Hz, the sub-genual chordotonal cells are maximally sensitive to vibration between 200 and 1000 Hz and also to sound below 2 kHz. As this is below the best range of the crista acustica it extends downwards the limit for airborne sound.

Hemiptera. The knowledge that many small Hemiptera communicate through substrate vibrations has led to the search for the receptor organs. In the pentatomid bug *Nezara viridula*, these consist of a system which is simpler than that in the Orthoptera but which operates in a similar manner (Cokl 1983). Recording from the leg nerves revealed a group of low-frequency receptors which responded phasically to upward deflections of the legs up to a frequency of 120 Hz. These potentials probably originated from the campaniform sensillae and are thus equivalent to type I receptors. Two further units, the middle- and high-frequency receptor neurons responded maximally to vibrations of 200 Hz and between 750 and 1000 Hz respectively. The sub-genual organ of *Nezara* possesses only two cell bodies to which these responses must be attributable. The parameters of the female calling songs are matched to the characteristics of these two receptors and, once again, we have an example of the close matching of a signal and its receptor system.

Arachnida. In spiders, the sense organs mainly involved in the reception of vibration are the slit sensillae, in particular those that make up the metatarsal lyriform organ. Slit sensillae are analogous to the campani-

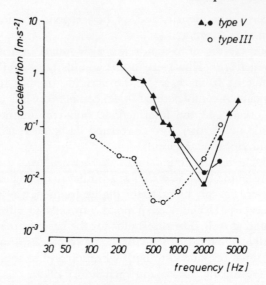

Figure 3.12. Threshold curves for a type III and two type V vibration receptors in the fore leg of the bush-cricket, *Decticus*. (From Kalmring, Lewis and Eichendorf 1978).

form sensillae of insects and comprise elongated pits in the cuticle. The bases of the depressions are covered by thin cuticle to which attach dendrites of the sensory cells; deformation of the cuticle stimulates the sensory endings. The metatarsal lyriform organ consists of a group of slit sensillae oriented across the dorsal surface of the metatarsus just above the tarsal–metatarsal joint. Upward movement of the leg compresses these sensillae, thus stimulating them. Distortions which produce lateral movement, however, also stimulate the receptors, and thus spiders can detect several different modes of vibration.

Barth and Geethabali (1982) recorded the neuronal activity of individual slit sensillae of the spider, *Cupiennuis salei*. If their sensitivity was measured in terms of acceleration, they were most sensitive to low frequencies with thresholds down to below $0.01 \, \text{mm s}^{-2}$. When their thresholds were expressed as displacement, however, sensitivity became greater with increasing frequency. Presumably this system allows spiders to discriminate between the complex vibrations due to signals of conspecifics and those of prey struggling in webs, both of which are made up of a wide range of frequencies and intensities.

In addition to the metatarsal lyriform organs, there are two slit sensillae on the claw of the proleg which respond both to airborne sound and to vibration. Being less sensitive than the lyriform organs, they extend the range of stimulus intensities to which spiders can respond. Nevertheless, ablation of both these types of receptor does not eliminate the spider's ability to detect vibrations, and it is highly probable that organs as yet undiscovered do exist (Speck and Barth 1982).

Crustacea. It is always difficult to identify the effective stimulus which elicits a response in sensory cells. This has been possible in, for example, the locust ear, where the mechanical properties of the tympanum and Müller's organ have been correlated with the stimulation of specific groups of scolopidia. One of the several difficulties in identifying the links in the stimulus–response chain is that the nature of the stimulus can change during the process. It is known that most substrates act as low-pass filters, and that vibrations are propagated dispersively. In addition, the nature of the vibrations may change as they pass from the substrate to the animal and then through the body to the site of sensory transduction. This difficult problem has been scarcely considered except in the Fiddler Crab, *Uca pugilator*, which uses vibratory signals in communication.

Crustacea, like the insects, have joint receptors which are responsive to substrate vibrations. Using laser vibrometry, Aicher, Markl, Masters and Kirschenlohr (1983) measured the vibration of different parts of the legs and carapace of fiddler crabs in response to vibrational signals. They found that the body of the crab acted as a differential filter, and that low-frequency components close to the repetition rates of the conspecific signals were amplified, while the high-frequency components were attenuated. Also interesting was the observation that the crabs modified their postures in response to drumming signals so as to maximise the signal-to-noise ratio, although the mechanism whereby they did this was not understood. It would not be surprising if such behaviour is common in those arthropods which use vibrational signals.

4. Neuroethology of Sound Reception

Behavioural assays of arthropod hearing have shown that a variety of temporal and spectral characteristics of acoustic signals can be recognized. In order to understand the behaviour associated with phonotaxis and species recognition more fully, it is useful to know how these signals are coded and processed by the nervous system. Most studies of sensory physiology have concentrated on the processing of intra-specific signals and, to a lesser extent, the use of sound in predator–prey relationships. This is understandable because of the relatively invariant nature of, for example, calling songs which have a clear function and provide an accessible starting point for analysis.

There is a tendency to consider arthropod acoustic receptors as having evolved to serve specific behavioural functions. Information which is irrelevant for these is filtered out and only the salient features required to initiate an appropriate response are retained. This may be true in some simple systems such as hearing in noctuid moths, but it should be remembered that many insects are capable of hearing a wide range of sounds. Some Orthoptera do not make use of acoustic signals but nevertheless have organs of hearing similar to those of their stridulating relatives.

One of the major aims of much recent research in this area has been to identify the recognition system for conspecific signals within the nervous system. The kinds of questions that have been addressed are as follows:

How is recognition of song carrier frequency accomplished? While peripheral filtering occurs to a variable extent, as described in the previous chapter, how is this process continued within the central nervous system?

How are the temporal features of song, which are so important for species identification, recognized? To what extent does pattern recognition occur within the thoracic ganglia as opposed to the brain, and what is the nature of this recognition system?

The different song parameters, carrier frequency, spectrum, temporal patterning, along with information from different sensory receptors, tympanal organs, vibration receptors and cerci, require to be integrated for an appropriate behavioural response. Where in the

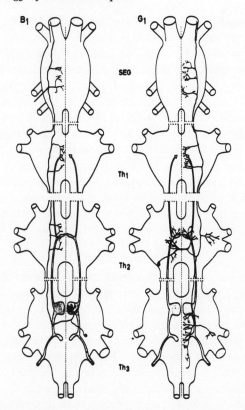

Figure 4.1. Two auditory ventral cord neurones in the locust. B1, ascending
neurone; G1, t-neurone. The shaded areas indicate the main auditory
pathways and neuropiles in the ganglia. SEG = sub-oesophageal ganglia;
TH_{1-3} = pro-, meso-, and metathoracic ganglia. (From Elsner and Popov
1978, after Rehbein).

nervous system does this occur and what are the mechansisms of
integration?

Phonotaxis requires that the direction of a sound source is gauged.
How is the binaural input processed to provide this information?

This work has been mainly carried out using acridids and gryllids but
cicadas have also received some attention. The examples described in the
previous chapter demonstrate that in almost all cases some degree of
stimulus filtering occurred, mainly with regard to carrier frequency.
This was due to the physical properties of the receptor system and of the
first order sensory cells. The next stage in the analysis is to see how this
information is further processed by the central nervous system (CNS).

Over the last ten years technical developments have enormously facilitated investigation of the functional morphology of the arthropod
nervous system. Histological agents such as horseradish peroxidase,

deoxyglucose and cobalt stains have revealed fine details of neuronal structure. The ability to make intracellular recordings from neurons and subsequently to inject dyes such as Lucifer yellow, which permeate to the ends of the axon and dendrites, has been crucial. Lucifer yellow fluoresces under ultraviolet light and mounts of the whole CNS can be examined. An entire cell with connections in the thoracic ganglia and brain, for example, can be visualized in this way. The relatively small number of nerve cells that appear to be involved in some of the responses is clearly an advantage, especially as the location of particular nerve cell bodies and their major dendritic and axonal processes is remarkably constant from individual to individual. This conservatism even extends between species. Neurones with similar physiological properties and morphology can be recognized from a range of different species. Valid comparisons can be made not only within, but also between species.

MORPHOLOGY

Neurones preferentially absorb cobalt ions and retain them within the cell membrane. Thus, if the cut end of a nerve is placed in a solution of $CoCl_2$, it will take up the cobalt which subsequently can be precipitated out as CoS which can be seen in either whole mounts or sections. This technique has been used to locate the position of the auditory neuropiles in several species of Orthoptera. These neuropiles are the regions of axonal and dendritic arborization, where the primary sensory nerves terminate and synapse with auditory interneurones.

Treating the auditory nerves of Ensifera in this way shows that the axons from the compound tibial sense organ of the front legs terminate in a well-defined ventro-medial region of the prothoracic ganglia (Eibl 1974). In Acridida, the main portion of the auditory neuropile is in the metathorax, although some of the sensory axons from the tympanic organ continue forwards to end up in the meso- or prothoracic ganglia (Figure 4.1) (Elsner and Popov 1978). In cicadas, the auditory neuropile lies in the fused abdominal-metathoracic ganglia (Figure 4.2) (Huber, Wohlers and Moore 1980). In all cases, as far as it is presently known, the primary sensory neurones terminate ipsilaterally, and any interactions between inputs from the two sides are mediated by interneurones.

The terminations of the sensory axons within the auditory neuropile are probably highly ordered, although this has been demonstrated clearly only in bush-crickets (Oldfield 1983). In these insects the tonotopic organization of the crista acustica is to some extent reflected in the central nervous system. High- and low-frequency neurones terminate respectively in the anterior and medial regions of the neuropile. In addition, in the species studied, *Caedicia simplex*, the terminal arborizations of the sensory cells tuned to the carrier frequency of the calling song of 16 kHz were much more extensive than those tuned to other

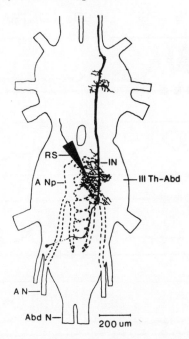

Figure 4.2. Auditory neuropile of the periodic cicada. A Np = neuropile;
AN = auditory nerve; III Th–Abd = fused metathoracic and abdominal
ganglia; RS = electrode with which recordings from the auditory
interneurone, IN, were made. Examples of recordings are shown in Figure 4.8.
(From Huber 1983).

frequencies. Some evidence indicating tonotopic organization within the
auditory neuropile exists for locusts (Rehbein, Kamring and Romer
1974) and crickets (Esch, Huber and Wohlers 1980) but these results are
not unequivocal as they are not based on the staining of individual units
having known physiological properties.

INFORMATION PROCESSING

Gryllidae

Thoracic auditory interneurones. Recordings from auditory interneurones within the prothoracic ganglia of crickets illustrate some of the
initial stages of information processing. Wohlers and Huber (1982)
describe six pairs of mirror-image interneurones with different physiological properties. All of these cells can be identified repeatedly in separate
preparations and homologous units have been found in other Orthoptera. The species used in this study, *Gryllus bimaculatus* and *G. campestris*, have calling songs with a carrier frequency of 4.5 kHz, and a specific
attempt was made to relate the characteristics of the interneurones to the
behaviourally relevant parameters of the song. The cells identified were

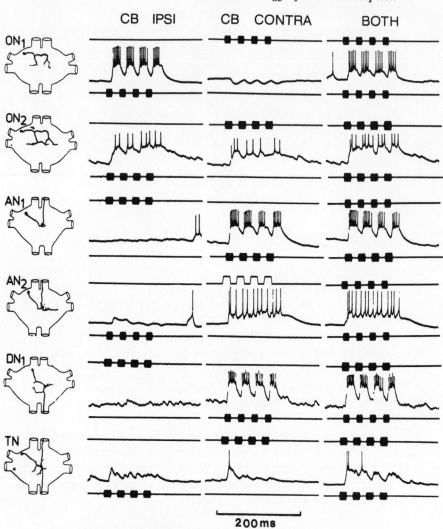

Figure 4.3. Auditory interneurones in the cricket. Responses of the different classes of neurone, middle channel, to simulated calling songs shown in the upper and lower traces, CB IPSI = sound produced at the side ipsilateral to the cell body; CB CONTRA = response to contralateral sound; BOTH = identical sounds presented simultaneously to both sides. For further explanation see the text. (From Wohlers and Huber 1982).

the omega neurones (ON1, ON2), so named on account of their shapes; they have terminal arborizations on both sides of the neuropile, ascending neurones (AN1, AN2), which send axons to the brain through the connectives contralateral to their cell bodies; a descending neurone (DN1), whose axon travels posteriorly and the T-neurone (TN), which sends branches of its axon to both the posterior thoracic ganglia and to the brain.

The responses of each of these classes of interneurone are summarised in Figure 4.3. The stimulus used was simulated calling song of four-syllable chirps having a carrier frequency of 4.5 kHz. By enclosing each fore leg in a small chamber containing a 1/4-inch loudspeaker, each tympanal organ could be stimulated separately or simultaneously. The neuronal responses of each cell type to sound contralateral and ipsilateral to the' cell body, and to simultaneous presentation, is shown in the central trace.

ON1 cells are excited by ipsilateral and inhibited by contralateral sound. Thus the trace in column 2 (CB CONTRA) shows hyperpolarising potentials due to summed inhibitory post-synaptic potentials (IPSP), while the firing rate when both ears are stimulated is less than when the ipsilateral one is stimulated alone (BOTH vs CB IPSI). The axonal branch which crosses to the contralateral side has the requisite morphology to mediate the inhibitory effect and the two ON1 neurones could form a reciprocal inhibitory pair. It is evident that an interaction of this type will greatly enhance the directional performance of the ears.

An additional property of ON1 cells is that they 'attend' exclusively to an ipsilateral sound if two sounds are presented at the same time. Even although both tympanal organs will respond to a sound regardless of its position if it is above threshold, the quieter input is inhibited by the louder one. This is effective only for a narrow range of frequencies around the carrier frequency of calling song. This means that if conspecific song is presented to one side of the animal and a different pattern of a different species for example, at the other, this will not interfere with the coding of the conspecific song. Thus song recognition and phonotaxis can occur even in an acoustically confused situation (Pollack 1986).

In contrast to ON1, the cross-ganglionic connections of ON2 neurones are excitatory, with the maximum response occurring when both ears are stimulated. These neurones do not code well for the temporal pattern of individual syllables.

AN1 interneurones respond only to an auditory input contralateral to the cell body, and this is expected from the position of its dendritic field. Individual syllables elicit bursts of spikes and a study of the frequency response of this cell shows that it is most sensitive to frequencies around 4.5 kHz.

The responses of AN2 cells vary between preparations, and Figure 4.3 shows a 'typical' pattern. Ipsilateral stimulation often results in low levels of excitation not usually leading to the production of spikes, while strong excitation is provided by contralateral stimulation. In some preparations, inhibitory interactions can also be observed. AN2 neurones are tuned to a broad range of frequencies. As these are above 10 kHz, AN2 cells are more suited to coding courtship rather than calling song. They would also respond to the high-frequency calls of hunting bats and might mediate avoidance behaviour. In addition, AN2 cells receive inhibitory inputs from vibration receptors in the leg (Kuhne, Silver and Lewis 1984) and it may be that this inhibition acts to switch between the two proposed behavioural functions of the neurones, court-ship song reception and predator avoidance.

DN1 cells copy the syllable structure of the chirps in the same way as AN1 although the best frequency response is around 2 kHz. However, their axons travel posteriorly to terminate in the meso- and metathoracic ganglia. This is one of a family of similar interneurones whose possible function is obscure. There are many interneurones in the mesothoracic ganglia of crickets which are known to respond to acoustic signals in a variety of ways (Elepfandt and Popov 1979) and it is likely that DN1 neurones synapse with some of these. Processing of auditory information at this level could allow rapid motor responses without mediation from higher centres in the CNS.

The final type of interneurone identified by Wohlers and Huber, TN, responds weakly to auditory stimuli. It is similar to AN2 as it also receives vibrational inputs. The two inputs appear to interact so that the neurone responds best to the higher-frequency courtship song pattern and excitation is enhanced at the repetition rates of the courtship song syllables (Kuhne, Silver and Lewis 1984). As courtship song is perfor-med at close quarters where vibrational stimuli are likely to be in range, it is possible that TN neurones are concerned in the coding of courtship song.

An additional type of ascending prothoracic neurone, AN3, whose morphology resembles that of AN1 and AN2, has been described by Boyd, Kuhne, Silver and Lewis (1984). Its physiological properties are similar to those of AN1 but it is an even better candidate for recognition of the calling song. Its threshold to 5 kHz sound pulses is around 15 dB lower than that of AN1 and it is more sharply tuned to the carrier frequency. This tuning is emphasized by inhibitory side-bands centred at 3 and 16 kHz.

It is clear both from the morphology and the physiological responses of these auditory interneurones that their functions are complex, and the results obtained to date do not reveal the full extent of their neuronal interactions. One ingenious way of investigating these interactions has been to make use of the property of the intracellular stain, Lucifer yellow. If a cell stained with this dye is irradiated with intense blue light from a krypton laser it becomes progressively inactivated over a period of a couple of minutes and then dies. This method can be used to record from and then selectively ablate individual, identified nerve cells.

Selverston, Kleindienst and Huber (1985) have used this technique to examine the relationship between the ON1 and AN2 interneurones. They recorded simultaneously from both ON1s and then observed the effect of photo-inactivating one of the cells on the behaviour of the other. They were able to confirm the direct reciprocal inhibitory connections between them; as one cell died the other was released from inhibition and a weak excitatory effect was revealed. The source of this latter effect was not identified but was probably mediated by other interneurones and, in the intact preparation, masked by the strong inhibition.

In a similar way they showed that the inhibition of AN2 by ipsilateral stimulation is also imposed by the activity of OM1. Further experiments of this type will undoubtedly reveal more about the detailed inter-relationships between the acoustic neurones.

AN1, AN2 and AN3 neurones project to the brain and recordings have been made from axonal arborizations in the proto- and deuterocere-brum which probably arise from some of these cells. One difficulty in interpreting some of the work on recording from neuronal processes in the thoracic ganglia and the brain is that the homologies both within and between species have not been clearly established. Much of this work was carried out over the same time period and different research workers have given homologous neurones different names. Thus, for example, AN2 of Wohlers and Huber (1982) is the same as their earlier A1AA (Wohlers and Huber 1978) and is likely to be equivalent to HF_1AN of Popov and Markovich (1982), int-1 in *Teleogryllus oceanicus* (Moiseff and Hoy 1982), ANA also in *T. oceanicus* (Hutchings and Lewis 1984) and PAHF1 (Boyan and Williams 1982). The homologies for AN1 and AN2 in gryllids have recently been clarified by the work of Hennig (1988).

It is clear, however, that the information coded by the thoracic auditory ascending interneurones becomes widely disseminated in the brain. In *G. bimaculatus*, the arborizations of PAHF1 (= AN2), which is the acronym for plurisegmental ascending high-frequency neurone number 1, are widespread within the brain and potentially synapse with central neurones involved in auditory and other behavioural responses (Figure 4.4) (Boyan and Williams 1982). The features of PAHF1, as recorded from within the brain, are those necessary for recognition of the courtship song. It responds over a range of between 5 and 100 kHz with a best frequency of 13 kHz, has a relatively high response threshold appropriate to the perception of nearby sound, and accurately copies the temporal pattern of courtship song with its short individual syllables.

The homologous unit in *Teleogryllus oceanicus* (int-1) has similar morphological and physiological properties to PAHF1 but, in this species of cricket, the calling and courtship songs have the same carrier frequency of 5 kHz and int-1 is therefore unlikely to be involved in

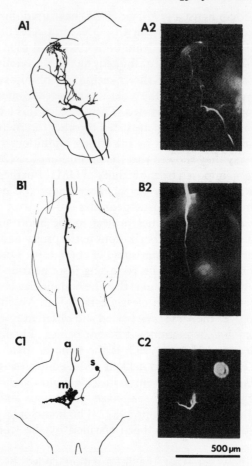

Figure 4.4. Anatomy of an auditory interneurone (Int1) in *Teleogryllus oceanicus*. On the left, camera lucida drawings derived from lucifer yellow stained preparation shown on the right. Lack of detail in the photomicrographs is due to the shallow depth of focus. The drawings were made by focusing at different levels within the ganglia. The cell body (s) is on the right side of the prothoracic ganglion (C), receives input through the dendritic arborization on the left side (m) and then send an axon (a) through the suboesophageal ganglion (B) to terminate in axonic branches in the brain (A). (From Moiseff and Hoy 1983).

courtship. Tethered flying *T. oceanicus* turn away from tones in the frequency range 15 to 40 kHz, and it is known that crickets respond with negative phonotaxis to the cries of bats. As int-1 responds over this frequency range, it has been suggested that the role of this interneurone is to mediate avoidance behaviour (Moiseff and Hoy 1983). Decapitated *T. oceanicus* will fly but no longer respond to ultrasonic pulses, which implicates the brain in the release of the escape behaviour. By means of the 'two-tone' paradigm in which sounds of two different frequencies are presented at the same time, Moiseff and Hoy showed that the excitatory response to a 30 kHz tone could be inhibited when presented along with a lower-frequency stimulus, and maximum inhibition occurred with a 5 kHz tone. This suggests a mechanism whereby the attention of the animal is locked onto the biologically important courtship stimuli, while other extraneous acoustic stimuli are inhibited.

Boyan and Williams (1982) also recorded the activity of another ascending neurone within the brain, PALF1, possibly homologous with AN1, which possessed the requisite characteristics enabling it to respond preferentially to conspecific calling song. It was sharply tuned to a best frequency of 5 kHz and, at this frequency, it followed the temporal patterning of the syllables precisely.

Brain interneurones. As well as recording from the terminations of thoracic acoustic interneurones in the brain, it has also proved possible to identify brain interneurones per se which process acoustic information. One such, PABN2 (pleurisegmental auditory brain neurone 2, Boyan 1981) is one of a pair of bilaterally symmetrical cells with extensive arborizations within the brain. These include the region, considered to be the auditory neuropile, which lies in the antero-ventral part of the protocerebrum where terminations of PALF1 and PAHF1 are also found.

Two-tone suppression experiments have shown that PABN2 has properties broadly similar to PAHF1 with a best frequency of between 13 and 16 kHz and maximum suppression of the response by a simultaneously presented 5 kHz tone. The magnitude of this suppression is dependent upon the relative intensities of the two tones, with total suppression occurring only if the 5 kHz tone is 15 dB louder than the 15 kHz one. Boyan argues that the calling song excites neurones, such as ON1, AN2 and AN3, which have best frequencies around 5 kHz, while at the same time suppressing high-frequency neurones like PABN2. Otherwise, high frequency neurones, which are tuned to the courtship song frequency and perhaps to bat cries, might be inappropriately stimulated by the high-frequency harmonics that are also found in the calling songs. In this way, possible confusion between calling and courtship songs, which elicit quite different behavioural responses, may be reduced.

Schildberger (1984, 1985) has recorded activity from a large number of brain neurones in the house cricket, *Acheta domestica*, and has classified neurones on the basis both of their morphology and physiological properties. One of the morphological classes, BNC1 (brain neurone class 1), has its arborizations in the protodeuterocerebral border where the ascending auditory neurones also terminate. Another class, the BNC2 neurones, connect, not with the ascending neurones, but with BNC1 units. DN cells (descending neurones) can receive inputs from both of the foregoing classes of brain cell and directly from the ascending neurones. Thus, on the basis of their anatomy, a number of sequential interactions are possible: AN → BNC1 → BNC2; AN → BNC1 → BNC2 → DN; AN → DN. The functional significance of these potential pathways is unclear, but obviously allows for complex interaction and integration of auditory information within the brain.

On the basis of their physiological properties the different brain neurones can be placed into two broad categories uncorrelated with morphology. These are, low-frequency neurones which respond maximally to tones around 5 kHz (the calling song carrier frequency in *A. domestica*), and high-frequency neurones. In general, the individual brain neurones are less specific in their responses than are the ascending auditory neurones. Thus, for example, none have been found which possess the sharp frequency tuning of ON1 or AN3. In addition, there are many neurones (GN = extrinsic glomerular neurones) which have very widespread connections in different regions of the brain. These can overlap those of all the other classes of brain cell as well as the ascending neurones. They also send branches to the mushroom and central bodies of the brain (Schildberger 1984). Such multimodal cells respond, not only to auditory stimuli, but to other sensory modalities such as vision and smell.

The duration of the syllables and their repetition rates in cricket songs are strictly patterned. These temporal parameters are important in eliciting female phonotaxis. The thoracic neurones, which are tuned to the calling song frequency, for example, AN3 and OM1, respond accurately to a wide range of sounds with differing proportions of syllables and inter-syllable-intervals, that is, with very varied mark to space ratios. Thus, while they code for the temporal parameters of the song, they do not preferentially filter the species-specific pattern. This filtering is likely to occur in the brain. Schildberger (1985) has demonstrated one of the ways in which this could happen. He mapped the responses of BNC1 and BNC2 neurones to 5 kHz syllables of varying lengths and intervals (Figure 4.5). Some of the BNC1 cells act as low-pass filters in not responding to high syllable repetition rates, others, as high-pass filters. BNC2 interneurones function as bandpass filters, responding only to syllable patterns within the range of the natural song. When the characteristics of these different filters are taken together they define a combination of parameters similar to those which elicit phonotaxis (Figure 4.5). There are therefore neurones in the brain, the BNC2 cells, which have the required properties for recognizing the temporal patterning of conspecific calling song. It is possible that BNC2 responses arise from an hierarchical pattern of interactions, AN1 → BNC1 → BNC2.

Vibrational interneurones. By comparison with tettigoniids and

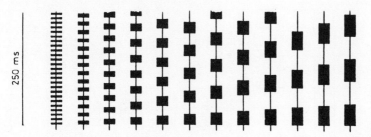

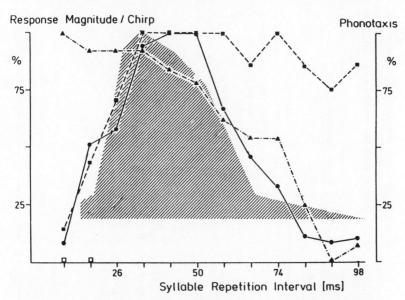

Figure 4.5. The responses of three different brain interneurones in crickets to simulated song pulses. The average energy of the simulated songs is kept constant, but the length of the individual syllables and intervals between them is varied systematically as illustrated above. A BNC1 neurone (■) behaves like a high-pass filter: two BNC2 neurones have the properties of a low-pass filter (▼) and a bandpass filter (●). The hatched area shows the relative effectiveness of the different simulated songs in eliciting phonotactic tracking. (From Schildberger 1984).

acridids, the processing of vibrational information in crickets has received rather little attention. The properties of two bimodal inter-neurones, TN1 and AN2, which respond both to acoustic and vibration-al stimuli, have been described above. OM1 is also influenced by vibra-tion but only at high stimulus intensities. The effect is inhibitory and the wide range of effective vibration frequencies suggests that its input is derived from both sub-genual and campaniform receptors. Inhibition of the auditory response reaches a peak at about 20 ms after the onset of the stimulus. When the animal is receiving both auditory and vibrational stimuli from calling song, this inhibition could have the effect of 'shar-

Table 4.1. The response characteristics of some classes of ventral cord interneurones in tettigoniids. VS neurones are those which are bimodal, responding to both acoustic and vibrational stimuli. S neurones respond only to airborne, and V neurones to substrate-borne stimulation. The classification is based on recordings from 228 neurones from the ventral nerve cord of *Decticus verrucivorus* and 104 from *Tettigonia cantans*. (From Kuhne, *et al.* 1984.)

Neuron type	Suprathreshold response characteristics to			
	Airborne-sound stimuli		Vibration stimuli	
	Response pattern	Response range	Response pattern	Response range
VS1	Phasic	7–100 kHz	Phasic to phase-locked	20–100 Hz
VS2	Phasic	4–100 kHz	Tonic to phasic	2–3000 Hz
VS3	Tonic	7–100 kHz	Tonic	20–5000 Hz
VS4	On-burst	4–100 kHz	On-burst	20–5000 Hz
VS5	Strongly phasic, suprathreshold responses only to combination of both stimuli			
S1	Phasic	4–100 kHz	–	–
S2	Phasic (quite similar to S1)	4–100 kHz	–	–
S3	Phasic on with tonic afterdischarges, long latency	7–30 kHz, no response to ultrasound	–	–
S4	Tonic	7–100 kHz	–	–
S5	On-burst	7–100 kHz	–	–
V1	–	–	Tonic	20–500 Hz
V2	–	–	Tonic	20–5000 Hz
V3	Unspecific responses	–	On-burst to tonic	20–5000 Hz
V4	–	–	Phase-locked	up to 200 Hz
V5	–	–	Phase-locked	up to 200 Hz

pening' the response of the ON1 neurone by suppressing neural activity during the inter-syllable intervals (Wiese 1981).

Units have also been described which receive inputs solely from vibration receptors in the legs. These can be divided into low-frequency (LF) units with best frequencies around 200 Hz, high-frequency (HF) ones which respond to vibrations of between 700 and 1000 Hz and medium-frequency (MF) units which have a flat response centred around 300 to 500 Hz. The latter are relatively insensitive compared to the first two (Table 4.1)(Kuhne *et al.* 1984).

HF interneurones also respond to airborne sound but, unlike TN1 and AN2, this is not due to convergent input from the tympanic organ but because the sub-genual organ also responds directly to sound.

The majority of vibratory interneurones receive inputs from all six legs, although that from one pair of legs is normally dominant. In general, the inputs are excitatory and additive in action. However, in LF3 the contralateral input is inhibitory and this suggests that the animal could discriminate the direction of a vibrational signal. This would depend on the attenuation of vibration being sufficient for a detectable difference to develop between the two legs, a distance of about 10 mm for a vibration occurring at right angles to the animal. *Gryllus* species live on the ground and soil is a sufficiently poor transmission medium for this to be a possibility. A cricket might be able to detect the position of a singing conspecific at close range by this means but it is more likely that the vibration detectors are not primarily involved in intraspecific communication but as part of the alarm system. This is not, however, necessarily true of other insects and, within the Orthoptera, tettigoniids, for example, certainly do make use of vibrational signals.

Cercal receptor interneurones. The individual syllables comprising the chirps in *Gryllus* species are repeated at 30 s^{-1} and this is an important recognition feature of the calling songs. The cercal hairs of crickets respond to low-frequency sound and should therefore be able to code for this syllable repetition rate (Palka, Levine and Schubiger 1977). Kamper (1985) has recorded from the axons of ascending interneurones located in the last abdominal ganglia which receive their inputs from the cercal hairs. While the major function of the cercal receptors is undoubtedly to mediate the escape response of the cricket, some of the interneurones show the requisite properties for near field sound reception.

Kamper found that when the hairs were stimulated with oscillating air movements at velocities of 2 cm s^{-1}, which is the intensity at which self-stimulation of the cricket by its own song occurs, some inter-neurones responded with phase locked potentials up to a frequency of between 100 and 200 Hz (Figure 4.6). The intensity of the stimulus was coded both by individual interneurones producing increasing numbers

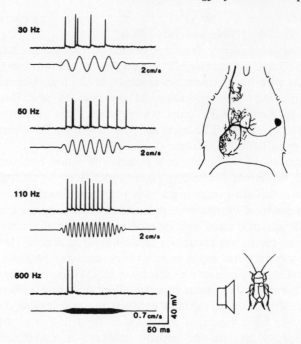

Figure 4.6. Responses of a cercal hair interneurone in *Gryllus* to low-frequency sound. Up to 110 Hz stimulating frequency the potentials are phase locked to the sound. At 500 Hz response to sound is almost absent. The anatomy of the interneurone in the last abdominal ganglion is shown on the right. (From Kamper 1985).

of potentials per cycle with increasing intensity and by recruitment of additional units having different thresholds.

As the cercal filiform hairs are particle displacement receptors, they are inherently capable of providing directional responses. It has been demonstrated that different groups of hairs respond preferentially to displacement in the longitudinal and transverse axes of the cerci. This is reflected in the behaviour of the cercal interneurones, each of which is maximally stimulated by air particle movement in a particular direction (Murphey, Palka and Hustert 1977). In addition, different neurones respond to specific phase angles of the sound. Central integration of these two parameters could provide accurate information concerning the direction of a low-frequency sound source.

While it has been demonstrated that the cercal receptor system does perceive the 30 Hz component of cricket song at a range of up to 15 cm, there is no behavioural evidence to show if this information is used (Kamper and Dambach 1981). However, the existence of interneurones which are tuned to 30 Hz does suggest that the cerci could be involved in courtship.

Recordings have also been made from the abdominal connectives of the East African Cricket, *Phaeophilacris spectrum*, to investigate the cercal responses to the very low-frequency air pulses that this species uses during courtship and aggression. These show that the air movements are accurately coded by the cerci. In this species, unlike the gryllids, there is no high-frequency stridulatory component of the wing display and the pattern of the syllables is very likely to be the only effective stimulus (Kamper and Dambach 1979).

Acrididae

Auditory interneurones. The investigation of acoustic processing by the central nervous system of crickets has been facilitated by the occurrence of calling songs which can be precisely characterized. In the species of *Gryllus*, *Teleogryllus* and *Acheta*, which are commonly used in experiments, they usually consist of syllables with strict temporal patterning and a sharp carrier frequency of 4.5 to 5 kHz. As the calling song is clearly a very relevant biological signal, there is good *a priori* reason to search for neural correlates of features of the songs, and this has proved a successful strategy. The songs of acridids (grasshoppers) and many tettigoniids (bush-crickets), by contrast, are produced by non-resonant systems and, consequently, have a wide-frequency spectrum. In addition, in many grasshoppers the temporal structure of sound pulses is blurred due to the asynchrony of the stridulatory leg movements. This makes the task of finding interneurones responsible for coding of the song more difficult.

Nevertheless, much of the earlier research into insect hearing was carried out with the locust and, to a lesser extent, tettigoniids. It was in the former that the location of the auditory neuropile in the brain was first identified and recorded from (Adam and Schwartzkopff 1967). It has subsequently been shown to be in the same region in other insects. Much of this work was carried out on the processing of auditory input in general and not specifically with the sounds used in communication, and I will concentrate only on the latter. A useful bibliography and review of the literature up to about 1976 is contained in Elsner and Popov (1978).

The bush-crickets and crickets are more closely related to each other than either of them is to the grasshoppers (Acrididae) and this is reflected in the structure and position of their tympanic organs. Nevertheless, because of the wide-band frequency content of acridid and tettigoniid songs and, possibly, their similar ecological requirements, the way in which they process acoustic stimuli has certain features in common. For example, while in crickets few of the thoracic auditory interneurones also receive inputs from the vibration receptors, in bush-crickets and grass-

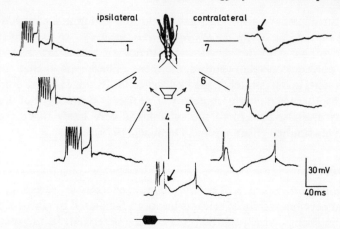

Figure 4.7. Directional responses in an auditory interneurone of the locust. Ipsilateral stimulation (1–3) give excitatory potentials with inhibition from contralateral stimulation (4–7). As the sound source is moved round the animal an inhibitory potential is seen first at 4 (arrow). The stimulus was a 20 ms, 20 kHz pulse 20 dB above the ipsilateral threshold. (From Romer *et al.* 1981).

hoppers the majority of such neurones are bimodal (Kuhne *et al.* 1984, Cokl, Kalmring and Wittig 1977).

There are about fifteen ascending interneurones in the locust which can be identified in recordings from the circumoesophageal commissures. It is possible to classify these on the basis of their physiological properties. For example, their responses may be phasic or tonic, they can have preferred frequency and/or intensity requirements and the majority possess directional features (Kalmring 1975). In general, their properties are more complex than those of the primary auditory neurones, and some signal features such as frequency response or intensity range are enhanced (Kalmring, 1978).

Because the acridid tympani are on the abdomen and not on the tibiae as in crickets, it is technically more difficult to stimulate each ear individually. This problem has been overcome by stimulating the tympani directly with piezoelectric mechanotransducers rather than with sound. Using this method it was possible to record from auditory interneurones in the metathorax of the locust and show that the majority of them received binaural inputs. Those from the contralateral tympanum could be either inhibitory or excitatory, and the integration of the two inputs with regard to the latency and magnitude of the response provides the potential for precise directional hearing (Figure 4.7)(Romer, Rheinlaender and Dronse 1981, Romer and Dronse 1982).

Most studies of locust hearing have employed either pure tones or pulses of white noise as acoustic stimuli. Modulating and patterning the latter to simulate the normal courtship sounds provides a more behaviourally relevant stimulus. Kalmring *et al.* (1978) showed that while none of the ascending auditory interneurones responded preferentially to the entire pattern of simulated song, different neurones coded for one or more of the appropriate parameters. There is therefore no peripheral filter for conspecific song, and the recognition mechanism must reside elsewhere within the CNS.

The possible role of the suboesophageal ganglion in processing sound has only recently been considered (Boyan and Altman 1985). However, in this ganglion there are ascending, descending and local interneurones which respond to auditory stimuli. The response latencies of all but one of these units are long, between 37 and 85 ms, which indicates that they are several steps along the path of stimulus processing. Some of the ascending units project to a slightly different area of the auditory neuropile than those from the thoracic ganglia. They are, in addition, bimodal, and also respond to optic stimuli. While the significance of these interneurones for signal processing is at present obscure, it is obvious that a straightforward hierarchical arrangement: auditory receptors → thoracic interneurones → brain interneurones, provides too simple a model.

Within the brain itself there are several cells which respond to the simulated courtship songs. Their properties are very similar to the bandpass neurones described in the cricket, in that, while they respond to a broad range of chirp repetition rates, their optimum response corresponds with that of the conspecific song. Unlike the ascending interneurones which transfer information from the ganglia to the brain, they do not code for the individual chirps, and they may therefore represent a higher level of integration within the system responsible for song recognition (Romer and Seikowsky 1985).

The morphology of some of the auditory interneurones in the thoracic ganglia of the locust has only recently been described (Romer and Marquart 1984). They identified several morphological classes within the metathoracic ganglia, including ascending interneurones (AN) and T-neurones with both ascending and descending processes which are homologous with some of those described from the neck connectives. In addition, there are local interneurones which make connections only within their ganglion of origin, segmental neurones (SN) and bisegmental neurones (BSN), which project to the mesothoracic ganglion. As might be expected, these respond in a more complex manner than the primary auditory neurones. Intracellular recording from these cells makes it clear that they play a complex role in processing the information which is transmitted to the brain.

Intracellular recording from interneurones in the grasshopper, *Omocestus viridulus*, some of which are clearly homologous with those of the locust described by Romer and Marquart (1984), have provided some intriguing data. In contrast to the static preparation usually employed,

Hedwig (1986) made his recordings with a grasshopper which was sufficiently unrestrained to be able to stridulate. Two ascending metathoracic interneurones produced bursts of potentials which were correlated with sound pulses of both its own and of simulated song. In this they resembled some of the ascending neurones of the locust. However, it became apparent that, during the period in which the animal was itself stridulating, these neurones were effectively 'deaf'. They were responding, not to the acoustic signal from the tympanal organs, but to a central and perhaps also proprioceptive input associated with stridulation. Even if the stridulatory movements produced no sound (because the file and scraper did not come into contact) the auditory neurones responded with bursts of potentials at the appropriate time. If external sounds provoked a response, they only did so at restricted phases of the signal. These interneurones are therefore involved, not only in processing auditory information, but also, seemingly, in some aspect of the motor control of stridulation.

Tettigoniidae

Auditory interneurones. Most of the interneurone recordings in tettigoniids have been carried out at the level of the ventral nerve cord and not from nerve fibres within the ganglia. The information transmitted to the brain has previously undergone a considerable amount of processing. In addition, much less is known about neuronal morphology than is the case in crickets and grasshoppers. In common with the latter, almost all of the tettigoniid ventral cord units respond to both airborne sound and to substrate vibration. They can be classified according to whether their responses are sub- or supra-threshold, that is, whether the relevant stimulus elicits propagated potentials (Kalmring and Kuhne 1980).

Approximately one third of the units recorded from in the neck connectives of two species of tettigoniid, *Tettigonia cantans* and *Decticus verrucivorus*, were found to be completely bimodal and responded equally to vibration and sound (VS units). The former stimulus was applied directly to the legs by means of a minivibrator, and the latter in the conventional manner with a loudspeaker. The remaining units responded predominantly to either sound (S units) or vibration (V units) but always with either excitatory or inhibitory subthreshold inputs from the other modality which modified their activity.

The three main categories of interneurone can be further subdivided, depending upon whether they are phasic or tonic, habituating or non-habituating. The majority of S and VS units were phasic and habituating, which meant that they responded to tettigoniid song for only a limited period and not for the entire duration of the song. The chirps or trills of tettigoniids can last for several seconds and contain hundreds of syllables. The duration, but not the temporal patterns of the chirps were coded by some tonic, non-habituating units.

Stimulating *T. cantans* with simulated calling song chirps, both with and without simultaneous presentation of vibrational stimuli, showed that the two sources could interact synergistically to enhance the response of certain VS interneurones. This had the effect of accurately coding for all of the syllables within a long chirp. However, such stimulation would only occur if the singing animal was close to the recipient, for example on the same shrub which could transmit the substrate vibrations associated with stridulation. At greater distances, the S and VS units alone could not provide the brain with such an accurate analysis of all of the song parameters (Kalmring, Kuhne and Lewis 1983).

Recording from single auditory units in the neck connectives of *Cicadicia simplex*, Oldfield and Hill (1983) also demonstrated the existence of different classes of interneurone possessing different frequency ranges and with tonic or phasic properties. One of these had an intermediate tonic–phasic response and was tuned to the 16 kHz carrier frequency of the calling song. The degree of tuning was sharper by a factor of two than that of the primary auditory fibres. Using the two-tone paradigm, they showed that the enhanced frequency tuning was due to lateral inhibition. All of the physiological properties of the interneurones could be derived in a relatively simple manner from the integration of excitatory and inhibitory inputs from the primary afferent auditory units.

Hemiptera: Cicadas

Auditory interneurones. In cicadas, as in Orthoptera such as the crickets, both the carrier frequency and the temporal patterning of song are important in intraspecific communication. The part played by ascending auditory interneurones within the thoracic-abdominal ganglionic complex has been looked at in two related sympatric species *Magicicada septendecim* and *M. cassini*. These two species can be found calling and courting in close proximity, and mechanisms of premating isolation are likely to be important. These cicadas are therefore promising subjects for looking at signal processing in the context of species recognition.

While recordings from the auditory nerves of both species show that each can 'hear' the song of the other and code for its temporal pattern, the interneurones of *M. septendecim* respond only to the conspecific song (Figure 4.8)(Huber *et al.* 1980). This is possible because, in this species, the spectrum of the calling song is fairly narrow and some filtering on the basis of frequency is possible. By contrast, the interneurones recorded from in *M. cassini* did code for heterospecific song, although not as accurately as for its own. In addition, they responded to broad-band 'clicks'. Presumably this lack of selectivity is a consequence of the

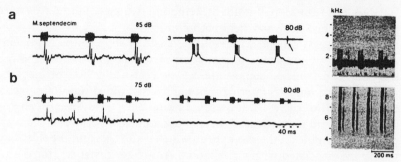

Figure 4.8. Auditory nerve responses of *Magicicada septendecim*, a, to its own calling song and b, to that of the sympatric cicada species, *M. cassini*. 1,2, auditory nerve responses; 3,4, auditory interneurone responses. Sound spectrograms of the songs of the two species are shown on the right. Heterospecific song elicits some activity in the auditory nerve but not from the interneurone. (From Huber 1983).

wide-frequency range of the song of this species which precludes the use of a frequency filter. While it is possible that the recognition system for conspecific song in *M. septendecim* could reside in the thoracic ganglia, this is less likely for *M. cassini*.

Crustacea: Fiddler crabs

Vibrational interneurones. The low-frequency vibratory signals of fiddler crabs of the genus *Uca* are detected by the joint receptors, and it is probable, but not proven, that these constitute the relevant receptor system. By recording from units within the leg nerves, Aicher and Tautz (1984) showed that *Uca* possess two classes of receptor. These were, wide-band units which are in the great majority and respond over a range of 2 to 2000 Hz with best frequencies of 15 to 30 Hz, and low-frequency ones which give an excitatory response betwen 2 and 100 Hz, and whose spontaneous activity is inhibited by vibration above the latter frequency. The source of this inhibition is unknown, but it means that the low-frequency units respond selectively to the range of frequencies found in the interspecific vibratory signals.

The way in which the central nervous system processes substrate-borne vibration signals, which presumably originate from such sensory units, has been investigated by Hall (1985a) in studies which are analogous to those in various orthopteran species. In *Uca*, possibly because of the absence of auditory input to complicate the picture, the response of the interneurones to vibration seems to be more straightforward. Hall identified five classes of interneurone on the basis of both their morphology and physiological properties (Figure 4.9).

The soma of all of the interneurones were situated within the posterio-medial region of the 5th thoracic ganglion, and each had arborizations within the five ipsilateral hemi-ganglia. These were connected to the

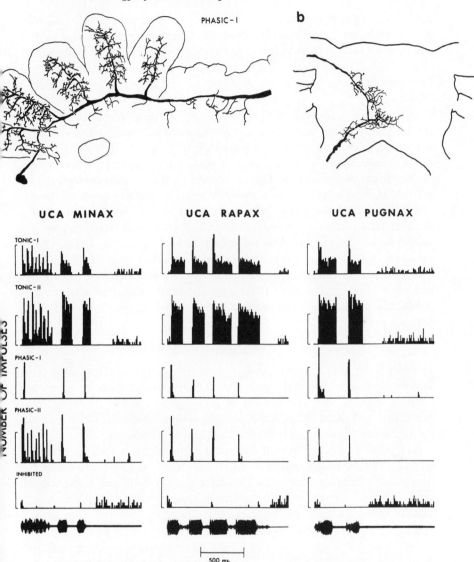

Figure 4.9. Vibration receptors in fiddler crabs. The morphology of a Phasic-1 interneurone in the abdominal hemi-ganglia (a) and the brain (b); c, responses of different classes of interneurone in three *Uca* species to conspecific vibratory signals. (From Hall 1985a).

anterior projecting axon which terminated in both the contralateral and ipsilateral areas of the tritocerebrum. The interneuronal classes could be separated on the basis of a number of anatomical features such as dimensions of the soma and details of the dendritic projections to the brain. A comparative study of several species of oxypodid crab, all of which make extensive use of vibrational signals, with a grapsid crab which is not known to do so, showed that the dorso-medial-tritocerebral neuropile was particularly developed in the former and could probably be considered analogous to the auditory neuropile of the insects (Hall 1985b).

Most of the interneurones had relatively flat response curves from 100 Hz up to 1 kHz, above which the threshold increased rapidly. The responses below 100 Hz were not investigated, and so it is difficult to relate these results directly to those of Aicher and Tautz. However, one of the classes of unit, Tonic-II, was excited by frequencies between 100 and 300 Hz and inhibited by higher frequencies, and so shows similarities with the low-frequency units in the leg. Characteristic responses of the different classes of interneurone to the vibrations produced by a signalling conspecific are illustrated in Figure 4.9c, and it is clear that both the duration and patterning of the pulses are adequately coded. Furthermore, the songs of other related species are equally well coded and there is no evidence of filtering for species-specific characteristics. Again, the recognition system must reside in the higher centres of the nervous system.

SUMMARY

In order to be able to make the appropriate behavioural response to an acoustic and/or vibrational signal the recipient needs to extract certain features from it. For recognition of a conspecific signal its temporal patterning and often the spectral composition are important. To locate the signal its direction and intensity require to be known.

As a general rule, the overall frequency response of an acoustic organ, measured by the compound nervous activity from the first order sensory neurones, is broadly tuned to the carrier frequency of the conspecific song. This is particularly true of those species such as crickets and cicadas with narrow-band songs. This selectivity depends, not only upon the properties of individual sensory neurones which go to make up the composite response, but also on the biomechanical and biophysical features of the auditory system. The antennae of *Anopheles* or the air sacs of female cicadas are structures which dictate, to a considerable extent, the responses of the animals to frequency. Even if the acoustic organ is able to respond to a wide range of frequencies there is often a concentration of units at the carrier frequency of song.

Within the ganglia are interneurones which are tuned to the carrier frequencies of conspecific song, and this tuning can be sharpened by the presence of inhibitory side-bands. The phenomenon of 'two-tone suppression', observed in some brain interneurones, ensures that inappropriate stimulation, arising from irrelevant sounds or potentially confusing harmonics, is inhibited. Most of the signal processing with respect to frequency appears to be carried out in the thoracic (and abdominal) ganglia and the resultant information is passed to the brain along relatively few channels.

In contrast to frequency coding, there is no evidence of selective filtering for temporal parameters of song at the level of the thoracic ganglia. Individual interneurones which respond specifically to the song carrier frequency normally code equally well for a wide range of syllable repetition rates. In the brains of crickets and locusts however, there are interneurones which, although they do not follow individual syllables, nevertheless are most sensitive to conspecific repetition rates. Such neurones could represent part of the recognition system for song.

Sound intensity is coded in the expected and conventional ways. Both the primary sensory units and auditory interneurones usually have normal sigmoid response curves. In addition, however, range fractionation occurs with different units having different intensity thresholds and ranges over which they can respond.

Direction of a sound source is coded by intensity differences at the auditory receptors. These differences are enhanced by a complex acoustic regime in some Orthoptera. They are designed to ensure that the differences are maximal at the conspecific carrier frequency, thus facilitating phonotaxis. These differences are amplified by the reciprocal inhibitory interactions that exist between contralateral inputs.

It should be emphasized that, just because units have been identified within the CNS which have the requisite properties to perform particular functions they must in fact do so. For example, the ON1 interneurone of crickets possesses the ability to recognise and code for the direction of conspecific calling song and, indeed it is reasonable to hypothesise that this is its function. Nevertheless, after photoinactivation of both ON1 cells, crickets are still capable of correct phonotaxis (Atkins, Ligman, Burghardt and Stout 1984). This indicates that the interactions between interneurones are complex and that alternative channels exist which are capable of providing the brain with the appropriate information.

Processing of acoustic and vibrational information in arthropods, in particular Orthoptera, is a very active research area at the present time. The properties of many of the building blocks, that is, the interneurones, have been described, while the more difficult task of investigating their complex interactions has started.

5. Neuroethology of Song Production

One of the early aims of ethologists was to describe the behaviour of animals in terms of motor patterns and their neural control. Insect song, particularly in the Orthoptera, provides a system almost ideally suited to achieve these aims. During singing, many insects perform highly stereotyped, repetitive patterns of behaviour which are good examples of 'fixed action patterns'. Additionally, they can often be elicited reliably, making experimentation easier. Other types of behaviour which have some of the requisite properties, such as different types of locomotion or respiration, have been the subject of successful neurophysiological investigation but are of much less interest to animal behaviourists. Consequently, neurophysiologists with behavioural leanings have attempted to investigate the motor control of song production, mainly in crickets and grasshoppers, but to a more limited extent also in other animals such as *Drosophila*.

One challenge to the neuroethologist was how to modify the traditional static preparation of the physiologist, so that recordings could be made from muscle and nerve in animals sufficiently unrestrained that they would still behave in a relatively normal manner. Several approaches to this problem have been developed. It is possible to implant thin, flexible wires into muscles and record electromyograms while leaving the animal relatively unrestrained. Alternatively, the animal can be held so that its legs are in contact with a polystyrene sphere which rotates as it walks. Sophisticated versions of this device have been built to measure phonotactic responses. This method has the advantage that precise electrophysiological measurements can be made. It has even proved possible to use suction electrodes and micro-electrodes to record from single fibres within the connectives or to make intracellular recordings from individual neurones. Regions of the brain or the connectives can also be stimulated to elicit stridulation. This provides not only information about the role of regions of the CNS in controlling behaviour, but a method to ensure performance of the behaviour.

One of the advantages of arthropods for this type of neuroethological experimentation resides in their neuromuscular physiology. Most of the muscles involved in stridulation in crickets and grasshoppers are fast muscles comprising very few motor units. Well-placed electrodes pick up clear electromyograms from single identifiable units, and it has been

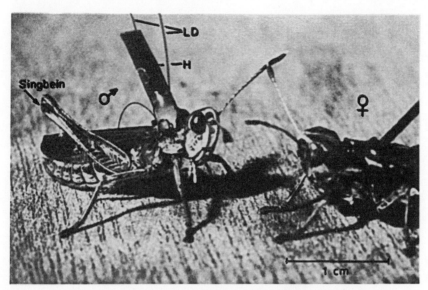

Figure 5.1. Courting *Gomphocerippus rufus* male with recording electrodes (LD) implanted into the leg muscles. Electromyograms from a preparation of this type are shown in Figure 5.6. (From Huber 1978).

possible to record from up to 16 simultaneously (Figure 5.1). In neurogenic muscle systems, which are universal in the Orthoptera, muscle potentials provide an accurate pattern of motor neurone output from the CNS as well as indicating the time of muscle contractions. It has thus been possible to obtain a fairly complete description of the sequence of neuromuscular events underlying the sound-producing movements of crickets, grasshoppers and some other insects.

The gross anatomy of the arthropod nervous system with its segmental arrangement of a brain and a chain of separate thoracic and abdominal ganglia, is a helpful feature. The different ganglia can be isolated easily by means of lesions and an initial indication of possible neural pathways and the location of control circuits deduced. The economy of neuronal units controlling both motor output and sensory processing means that it might be possible eventually to describe the central control of acoustic behaviour in terms of interactions and connections of individual nerve cells. Unfortunately, the apparent simplicity resulting from the involvement of small numbers of units is at least partially offset by the fact that different dendritic and axonal arborizations of the same cell may possess different information processing properties. It is not therefore sufficient to record from the cell body alone.

The more important questions that have been addressed by neurophysiologists interested in the motor control of acoustic behaviour are:

– The location of the command centres controlling song

- The location of the pattern generators for song
- The circuitry of the pattern generators
- Initiation and maintenance of motor output
- Co-ordination and switching between different patterns and phases.

STRIDULATION IN GRYLLIDS

Site of controlling centres

The earliest investigation of the neurophysiological basis of song production in insects and the one which led directly to much of the subsequent research was that of Huber (1955, 1960). By stimulating different areas of the brain of the cricket, *Gryllus campestris*, particularly those in the region of the mushroom and central bodies, he was able to elicit all three song patterns: calling, courtship and rivalry. These results were extended by Otto (1971) who was able to implant flexible, thin, stimulating electrodes into the brains of freely moving rather than tethered animals.

The initial hope of these experiments was that localized song centres in the brain controlling the expression of the different song patterns would be identified. However, this was not realised, as very few locations were found from which it was possible to elicit only one song type (Figure 5.2). In general, depending upon the frequency and intensity of the stimulating pulses, two, or in some cases all three song patterns, could be obtained from the same locality. Calling song usually had the lowest threshold of activation. Inhibitory areas were also identified where stimulation switched off ongoing song.

Crickets with their neck connectives severed remain silent for a period following the operation but, after a few days, spontaneous singing occurs. Calling song predominates but some of the elements of courtship and rivalry are also observed (Kutsch and Otto 1972). Non-patterned electrical stimulation of the cut connectives can also elicit long periods of stridulation (Otto 1967). It is clear from these experiments that the brain must be involved in integrating sensory information important for initiating song and in choosing the behaviourally appropriate type. However, the neuronal circuitry for generating the cyclical neuromuscular events producing the stridulatory wing movements must reside in the thoracic ganglia. As normal song can also be produced with the nerve cord cut betwen the meso- and metathoracic ganglia, the location of the circuitry can be further narrowed to the first two thoracic ganglia.

The concept of the 'command' interneurone was first proposed for Crustacea in which it was demonstrated that the activity of individual fibres could switch on complex patterns of neuronal activity, controlling behaviours such as swimming and escape. It was suggested that brain stimulation in crickets was, either directly or indirectly, also stimulating

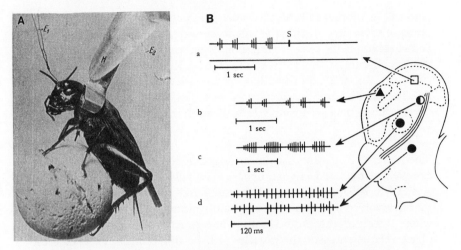

Figure 5.2. A, Male *Gryllus bimaculatus* suspended by a holder (H) and grasping a light cork sphere which rotates as it walks. An indifferent electrode, E_2, is attached to the prothorax and the stimulating electrode, E_1, is implanted into the brain; B, Longitudinal section of the cricket hemi-brain showing the effects of electrical stimulation of different regions. Stimulating the region marked □ resulted in inhibition of ongoing courtship song as illustrated in a. S marks the onset of stimulation. Stimulation of the regions ▲, ◖ and ● released calling (b), aggressive (c) and courtship (d) songs respectively. (From Huber 1960).

command interneurones for stridulation. Bentley (1977) was able to elicit calling song in crickets by stimulating at most a small number and possibly only a single axon in the neck connective. This axon is located in the same region of the connectives of different individuals of the same species and also in members of different genera (*G. campestris* and *Teleogryllus oceanicus*). In both these species increasing the stimulus frequency increases the chirp rate, although the pattern and timing of events within individual chirps remains unaffected. In *T. oceanicus* the rate of stimulation of the 'command' axon also controlled which song pattern was performed; medium rates of stimulation (50 to 70 s^{-1}) resulted in calling song, high rate ($> 100 s^{-1}$) in rivalry, while a pattern similar to that of courtship was produced by prolonged stimulation at low rates (30 to 40 s^{-1}). It appears therefore that there are not several specific command interneurones each responsible for eliciting a single song type, but rather that selection depends upon the intensity of the activity in possibly a single channel.

However, the foregoing almost certainly represents an oversimplification because Bentley was unable to elicit entirely normal courtship song by stimulation of the connectives. In addition, Otto and Weber (1982) have recorded from descending neurones in the connectives whose patterns of discharge reflect various parameters of the song patterns. Some of these fired during the 'silent' period of song and one was active before each syllable of courtship song. Nevertheless the function of these units is obscure. Thus, while the circuitry within the thoracic ganglia is sufficient for the co-ordination of the motor control of the wing movements and for the pattern of syllables within chirps, the chirp rhythm itself may be strongly influenced by the brain.

Control of motor patterns

The patterns of neuromuscular activity underlying cricket stridulation as revealed by electromyograms are relatively straightforward (Ewing and Hoyle 1965, Bentley and Kutsch 1966). The muscles of the meso-thorax involved in stridulation are mainly bifunctional and also move the wings during flight. Additionally some function as leg muscles. The phase relationships of the muscles will be different for each of these activities, in particular between walking on the one hand and flight and singing on the other. Muscles which are synergistic in action during the former can be antagonistic when moving the wings. This multi-functional role for the muscles has been well documented in locusts, and it implies that the neuronal circuitry which controls these behaviours has considerable flexibility and is not rigid or 'hard wired' (Figure 5.3)(Wilson 1962). It has been shown in crickets that almost all possible phase relationships occur in the patterns of firing in muscles which nevertheless possess a fixed phase during song (Elephandt 1980).

The action of the wings which gives rise to song in crickets is an opening and closing movement of the raised elytra, with sound pulses usually confined to the closing phase. Wing closing is analogous to the upstroke during flight, and simultaneous sound and muscle potential recordings show that the main indirect flight muscles responsible for upstroke in Orthoptera, the dorso-ventrals, are active just before each sound pulse. The antagonistic, direct downstroke muscles, the subalar and basalars, are active during the inter-pulse-interval and contribute to opening the wings. The indirect depressors, the dorsal longitudinal muscles which produce much of the power for flight, are relatively unimportant in stridulation. They do not appear to be involved in the calling and aggressive songs but do fire during the loud sound pulses or 'ticks' that occur in the courtship song. The ticks are associated with larger wing excursions than are the quiet sound pulses which comprise most of the courtship song (Bentley and Kutsch 1966).

The major variable between individual sound pulses or syllables which make up the chirps is their intensity. The intensity of the syllables is increased by the recruitment of additional muscle units within muscles and by multiple firing of the muscle units which also occurs during flight. It is noteworthy that the carrier frequency remains constant and

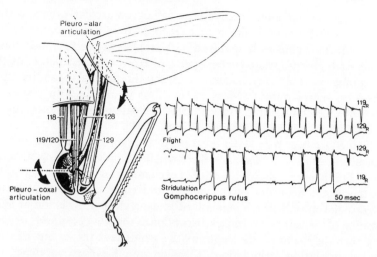

Figure 5.3. Bifunctionality of metathoracic leg and wing muscles in the grasshopper *Gomphocerippus rufus*. During flight, muscles 119/120 and 129 (also 118 and 128) act antagonistically so as to move the wing up and down. During singing they fire synergistically so that the wing does not move but the leg performs stridulatory movements. (From Elsner 1983).

is independent of intensity, which means that, although the power developed can change, the velocity of the closing stroke stays the same. This is because the velocity of wing movement is regulated by the mechanical properties of the file and plectrum and the resonance of the harp. Elliott and Koch (1985) have coined the phrase, the 'clockwork cricket', because of the analogies with the working of a clock. The harp is equivalent to the pendulum and sets the frequency, muscle power is equivalent to the weights or spring, while the file and plectrum together function as the escapement.

Stridulation in crickets resembles behaviours such as respiration, flight and locomotion in which sets of antagonistic and synergistic muscles contract with specific phase relationships, resulting in the performance of repetitive or cyclical patterns of activity. The important and interesting question concerns the nature of the neuronal circuitry within the nervous system which produces this type of output to the muscles.

The first factor which needs to be evaluated is the possible role of sensory input or feedback in generating the output pattern. The experiments described earlier which involved transecting the nerve cord at various levels demonstrate that the motor pattern is generated within the meso- and metathoracic ganglia. Sensory input is normally important in initiating stridulation, in selecting the song type and affecting its tempo. It may originate in the brain or the tympanal organs or mechanoreceptors, signalling the presence of a mature spermatophore. These sensory influences do not affect the pattern of neuromuscular events within chirps. A further possibility is that feedback from sense organs within the thoracic segments is involved and proprioceptive feedback could generate the appropriate cyclical pattern. However, destruction of the wing hinge stretch receptors and the nearby chordotonal organs has little effect on motor output (Moss 1971). This is in line with results of investigations on flight in locusts where the patterning of motor output is unaffected by deafferentation (Wilson 1966).

The elytra bear hair plates adjacent to the plectrum and file on both dorsal and ventral surfaces, and these will be stimulated during stridulation. This has been shown to result in excitatory post-synaptic potentials in the motor neurones of the opener and closer muscles and in the pleuro-alar muscle which is involved in wing closing. Ablation or inactivation of the hair plates causes disruption of the wing position, suggesting that they control the angle of the wings and the extent of their excursion during stridulation. They do not, however, influence the pattern of motor output (Elliott 1983).

Most recently Schaffner and Koch (1987) have looked at the role of campaniform sensillae which lie on the lower surface of the cubital vein. These are found only in males of the *Gryllus* species and therefore are likely to have a role in stridulation as the females do not sing. Cutting the cubital nerve, thus eliminating the sensory input from the sensillae to the thoracic ganglia, had profound effects on song pattern. Syllable durations became variable, and on average of shorter duration, while the timing of the syllables was distorted. Nevertheless, electromyograms from the opener and closer muscles of operated crickets demonstrated that, in spite of the atypical song, the pattern of motor output during stridulation remained unaffected. Deafferentation of the sensillae affects the juxtaposition of the file and scraper. In relation to the clockwork analogy, the escapement has become deranged but the pendulum is unaffected.

The neurnal circuitry controlling several cyclical or repetitive behaviours in arthropods is understood in principle if not in detail. However, in spite of the length of time since the first attempts to elucidate the control of stridulation in crickets, it is rather less well understood than is, for example the neuronal mechanisms underlying locust flight or crustacean trituration. The first model was proposed by Ewing and Hoyle (1965) and, although heuristically useful, is unlikely to be correct. It is based on the same principle put forward by Wilson (1964) to account for the motor output to the flight muscles in locusts, and it was parsimonious to suggest a common mechanism for two rather similar behaviours. The model states that the motor neurones innervating the antagonistic sets of opener and closer muscles possess, either directly or indirectly via interneurones, reciprocal inhibitory cross-connections,

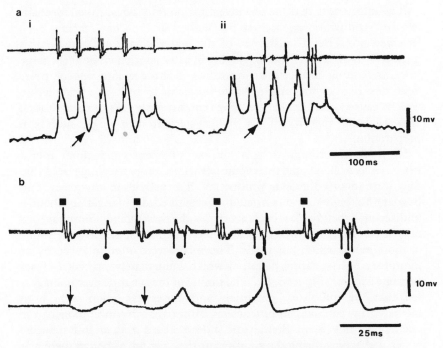

Figure 5.4. a, Second basalar motor neurone activity during chirping. Lower trace; intracellular activity, the arrows indicate inhibitory post-synaptic potentials. Upper trace; concurrent activity in i, the second basalar motor nerve and ii, the first promotor nerve; b, Upper trace: extracellular activity in the second basalar nerve (■) and the first promotor nerve (●). Lower trace: Intracellular record from a closer (first promotor) neurone. Closed arrows, onset of IPSPs. (From Bentley 1969).

and that they are driven by a common input. Such a model has the requisite properties and has been proposed for a number of alternating output systems including the production of tettigoniid song (Josephson and Halverson 1971).

It is not possible to test the validity of such models without first making intracellular recordings from the motor neurones so as to demonstrate the actual interactions between the neuronal units. This was done initially by Burrows (1973) and more recently by Robertson and Pearson (1985) investigating flight in the locust. They have shown that flight is controlled, not by a network based upon reciprocal inhibitory connections, but mainly by excitatory interactions modulated by delay circuits. The results of some of the recent research on cyclical behaviours in arthropods suggests that non-spiking interneurones are involved as pacemakers for setting the oscillatory frequency. There are, as yet, no reports of such interneurones influencing the timing of the events during stridulation.

It is unfortunate that the one investigation into the neuronal mechanisms of stridulation in crickets by Bentley in 1969 has scarcely been followed up. The intransigence of the material may account for this neglect. It is difficult to induce singing in a preparation while at the same time implanting intracellular recording electrodes into one or more neurones. Bentley was able to accomplish this by making a brain lesion which caused the cricket to sing continuously for several hours. However, the majority of his results stem from a single successful preparation.

He made recordings from a number of opener and closer motor neurones as well as some interneurones whose properties suggested that they were involved in song production. The activity in an opener (2nd basalar) motor neurone consisted of a marked depolarization about 3 milliseconds before the occurrence of the first spike. Simultaneous muscle recordings showed that each motor neurone spike gave rise to one or more motor action potentials. The spikes were often followed by an inhibitory post-synaptic potential which immediately preceded closer neurone activity (Figure 5.4a). The timing of the synaptic events suggested that the inhibitory potentials and closer excitation were both mediated by common interneurones rather than by inhibition of the openers by the closers. Bentley did indeed record from an interneurone which was active during the appropriate time period, although there was no direct evidence that the unit was involved in this interaction.

Activity in the closer motor neurones started with an inhibitory post-synaptic potential between 3 and 7 milliseconds after the first opener spike, followed by a depolarization of the neurone. Each successive closer motor neurone depolarization was associated with a burst of action potentials in the closer muscles. These were closely phase locked to the opener motor neurone spikes with a latency of 16 milliseconds (Figure 5.4b).

On the basis of these results Bentley has proposed a model for generation of the syllable pattern. The cycle is initiated by the opener motor neurones receiving an excitatory input from a slow oscillator, the chirp generator. Positive feedback between opener motor neurones results in the production of synchronised bursts of potentials. A similar mechanism for neuronal bursting has been shown to occur in neurones implicated in the escape response of the nudibranch mollusc, *Tritonia* (Willows 1968). The opener neurones, via interneurones, feedback on themselves so as to inhibit their own activity and thus terminate the burst, while at the same time inhibiting closer activity. This inhibition is lifted when the opener burst ceases and the closer motor neurones start firing, possibly due to post-inhibitory rebound or to direct excitation. This model is consistent with the experimental data and accounts for the latency locking found between opener and closer activity. It also has features in common with the model of the flight motor proposed by Robertson and Pearson (1985) which is not unexpected, as flight and song share many motor units. However, this is a topic which would benefit from additional research.

The model described above only concerns the syllable rhythm generator and we have little idea at present how the chirp pattern is produced. Bentley did record activity in interneurones which might belong to this system. Some interneurones were active during the inter-chirp interval and others during the chirps. Potentials were not correlated with individual sound pulses, suggesting that the interneurones were part of the chirp pattern and not the syllable pattern generator. It is not clear what relationship these neurones bear to those with similar properties in the neck connectives.

STRIDULATION IN ACRIDIDS

The experiments which were carried out on the control of cricket song stemmed from the early interests of Franz Huber, who also provided the initial inspiration for a parallel study on song in gomphocerine grasshoppers by Norbert Elsner and his co-workers. These workers have concentrated on two species, *Gomphocerippus* (= *Gomphocerus*) *rufus* and *Omocestus viridulus*. Both stridulate by scraping a file on the inner side of the femur of each hind leg across an elytral plectrum, and the mechanism and types of sound produced are described more fully in Chapter 2.

The analysis of the control of stridulation in grasshoppers has not advanced as far as in crickets. Recordings have not yet been made either from the motor neurones which drive the relevant muscles or from any interneurones which are likely to be involved in generating the output patterns. However, there are a number of particularly interesting features of grasshopper song which are absent in crickets.

For example, in *G. rufus*, stridulation is only one of several motor acts which together make up the courtship sequence performed by males. These other movements involve the head, fore body, antennae and appendages and it is revealing to examine not only the control of stridulation, but its relationship to the other components. The stridulatory movements themselves change systematically throughout the sequence and there is the added complication that the two legs execute different patterns. Grasshopper courtship thus provides the opportunity to examine not only the control of stridulation but also the transition from one element of behaviour in a sequence to another.

The courtship display of *G. rufus* consists of three phases or subsequences. The sounds associated with the display along with the stridulatory leg movements recorded by means of a Hall effect generator are illustrated in Figure 2.15. The first subsequence commences with sideways movements of the head oscillating between once and twice a

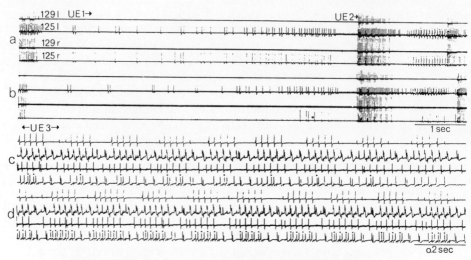

Figure 5.5. The activity of the subalar (129) and pleurocoxal (125) muscles
on the two sides of the body during stridulation in *Gomphocerippus rufus*.
Four consecutive traces a to d are illustrated. The three sub-sections of the
display are labelled UE1, UE2 and UE3. (From Elsner and Huber 1969).

second. At the same time, the rear legs move up and down at double the
head-shaking frequency, and each downward leg stroke produces a weak
sound pulse. Towards the end of this subsequence the tempo of the
movements increases to about 8 to 10 per second. In the second subsequence, the grasshopper raises the anterior part of its body and jerks its
antennae first up and back, then down and back up again, while the hind
femorae are raised rapidly a few times as it kicks its tibiae back. These
femoral movements are associated with loud sound pulses. The third
subsequence comprises a series of around 40 chirps, each of 6 to 8
syllables or pulses. Muscle potentials recorded during this subsequence
are shown in Figure 5.5 (Elsner 1973).

Initiation of courtship and the role of sensory input

As in crickets, sensory input is important in the initiation of courtship,
and the visual stimuli of an approaching female is probably the most
usual. However, no specific stimulus is essential, and covering the eyes
and/or cutting the tympanal nerve does not entirely inhibit courtship
and stridulation by the male and does not affect the pattern of song
production. Likewise, abolition of proprioceptive feedback from the
hind legs by amputation has little effect either on the motor output to the
leg muscles or on the overall courtship pattern. In such mute amputees
the normal courtship sequence is followed, with a silent period corresponding to the time when stridulation would normally have occurred.
Electromyograms of the leg muscles, which are in the metathorax, show
normally patterned activity, but at a lower level than usual.

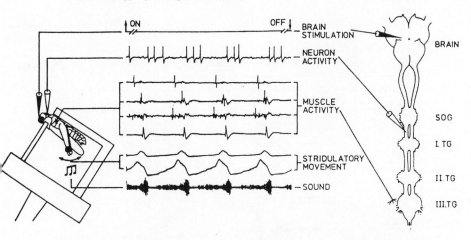

Figure 5.6. Preparation in which brain stimulation is used to elicit song in the grasshopper, *Omocestus viridulus*. Song, leg movement, neuronal activity and electromyograms can be recorded simultaneously. SOG = sub-oesophageal ganglion; I-, II-, III-. TG = pro-, meso-, meta-thoracic ganglia. (From Hedwig and Elsner 1985).

One operation which does have a major effect on stridulation is immobilization of the head, so that the animal cannot carry out those parts of the first subsequence that involve head rocking. If this is done, the remainder of the behavioural subsequences are totally inhibited. This suggests that the overall pattern is organized in a sequential or hierarchical fashion whereby the performance of the head movements triggers or releases the subsequent behaviours.

These conclusions are reinforced by experiments in which the nerve cord is cut at various levels. As long as one of the two connectives is complete, the entire courtship sequence is performed, although there may be some initial short-term disruption following the operation. Descending command neurones on either side of the body must therefore be able to trigger the appropriate circuitry within the ganglia. However, transecting the cords results in the appearance of only those behaviours carried out by structures anterior to the cut. Thus, if the ventral nerve cord is cut between the second and third thoracic ganglia, the grasshopper makes the normal head and antennal movements but no stridulatory movements occur. However, after the appropriate period for stridulation has elapsed, the sequence repeats, starting with subsequence one. Thus the overall co-ordination and timing of the courtship sequence is controlled by the anterior part of the central nervous system, probably the brain. It is possible, as in crickets, to electrically stimulate the brain and so elicit the entire sequence of courtship behaviour (Wadepuhl 1983).

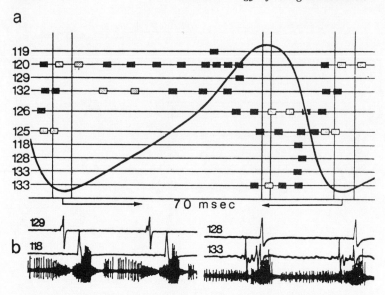

Figure 5.7. a, A motor score for a single chirp of *O. viridulus*. The solid line indicates movement of one leg and the blocks, motor unit activity; b, Examples of activity in four of the muscles along with simultaneous sound recordings on the lower trace. The muscles are numbered after Snodgrass, 1929. (From Hedwig and Elser 1980).

The motor programme underlying both stridulation and the other components of courtship behaviour has been worked out in both *G. rufus* and *O. viridulus* by means of electrodes implanted into the muscles (Elsner 1973, Hedwig and Elsner 1980). This technique has been developed to a very sophisticated level in these preparations, with more than sixteen electrodes being implanted simultaneously. In addition, the grasshopper can be restrained if necessary and made to stridulate by electrical brain stimulation (Figure 5.6). This has allowed a fairly complete picture of the activity in the muscles controlling movement of both legs to be constructed, and also of the behaviour of the different motor units which make up the individual muscles.

The individual leg muscles are composed of two or more motor units, each innervated by a separate fast motor nerve whose activity gives rise to a phasic contraction. The majority of these units are also innervated by slow axons which produce tonic muscle responses. The fast stridulatory movements are associated primarily with the former type. Within a muscle, a consistent pattern of firing is always seen, with one of the units having a lower threshold, while another may be recruited during the production of larger or more powerful leg movements. Thus the overall pattern of motor activity is complex, with different muscles and also the separate units within muscles, having preferred times of firing. In general, the two sets of antagonist muscles concerned with the up- and downstroke of the leg fire out of phase with each other. However, even within synergists there is a fixed sequence of activity and the muscles do not fire in synchrony. A so-called 'motor tape' for a stridulatory movement in *O. viridulus* is illustrated in Figure 5.7.

Unlike the pattern in crickets the bilateral synchrony changes during the courtship sequence. During subsequence 1 of *G. rufus*, the two legs move in synchrony, while during the remainder of the sequence they move out of phase with one leg leading by about 6 milliseconds. This lead is periodically transferred between the two sides. Peripheral input is implicated in this alternation; if one leg is amputated or fixed, the neuromuscular pattern is not changed but the alternation between legs is abolished.

In *O. viridulus* the pattern is yet more complex as the movements of the two hind legs, and therefore the sounds that they produce, are different, and thus two distinct motor tapes are played at the same time. Alternation of pattern between the two legs is also seen in this species. This is a further demonstration of the flexible nature of the connections between motor neurones whose output can switch between different modes, not only when changing between different behaviours such as flight and stridulation, but also during the course of their performance.

While the work of Elsner, Huber and others in this area has provided us with a detailed picture of the neuromuscular events underlying stridulation in grasshoppers and has also demonstrated its essentially endogenous control, we have little idea of the neuronal circuitry which gives rise to the output. Even the role of peripheral feedback is not clear. Thus, while amputation of one or both hind legs has little immediate effect on output patterns, over a period of hours or days, it leads to considerable changes in the co-ordination of motor output, and sets of muscles which are normally antagonistic during song, may even fire in synchrony. A complete model for stridulation will eventually have to take into account not only immediate influences on the system but also such longer-term effects.

CONTROL OF SONG IN DROSOPHILA

Because of their size, *Drosophila* would appear to be an unpromising subject for studying the neurophysiological basis of sound production. While it is certainly true that there are considerable problems in recording from units within the central nervous system of singing flies, this disadvantage is offset by the possession of positive features not found in other insect groups. The intensive use of *Drosophila* species in genetical research over a period of more than seventy years has laid the foundation for several novel approaches to neurobiological problems. One such is

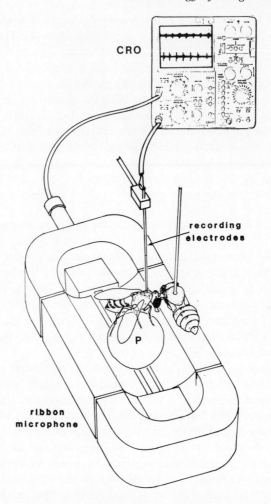

CRO

recording
electrodes

P

ribbon
microphone

Figure 5.8. Preparation for the simultaneous recording of song and muscle potentials in *Drosophila*. The male is suspended by electrodes implanted into the flight muscles and grasps a polystyrene bead (P). The female, with wings and legs removed, is presented to him suspended from a micromanipulator. Sound is picked up with a ribbon microphone and song and muscle activity are displayed on an oscilloscope.

the ability to manipulate the behaviour under investigation by selection or mutation.

D. melanogaster, the species used for the majority of genetical and behavioural research, performs two types of song during courtship – sine song and pulse song. The former results from low-amplitude wing vibrations of one extended or partially extended wing. Pulse song is produced by single up-and-down wing movements of about three mill-

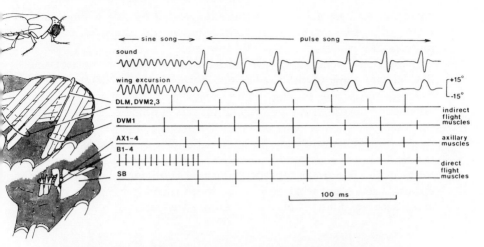

Figure 5.9. Summary diagram to show the neuromuscular and mechanical events associated with courtship song production in *Drosophila melanogaster*. Vertical lines represent muscle potentials. DLM = dorsal longitudinal muscles; DVM = dorsoventral muscles; AX = axillary muscles; B = basalar muscles; SB = sternobasalar muscle. (From Ewing 1979a).

iseconds duration, repeated at thirty per second (Bennet-Clark and Ewing 1968).

It is possible to implant fine, insulated wire electrodes into the wing muscles of tethered, singing male *Drosophila* and to record their electromyograms. The method used is illustrated in Figure 5.8. A similar technique was used to record from muscles during flight, and this work has resulted in a model for the generation of the flight motor pattern (Wyman 1973). A male *Drosophila* can be induced to sing by presenting females, and then recordings of muscle potentials and song are made simultaneously.

The neuromuscular system of Diptera differs from that of Orthoptera in some important respects. The muscles which move the wings in Diptera are divided into two categories, the indirect power muscles, often termed myogenic or fibrillar muscles, and the so-called direct and axillary neurogenic or non-fibrillar muscles. The functions of the latter are to extend and fold the wings and to modify various wing beat parameters. While these non-fibrillar muscles function in a conventional way and contract in response to motor action potentials, the fibrillar muscles are stretch activated. In other words, action potentials make the muscles competent to contract, but actual contraction results from stretching the muscles. Thus the potentials recorded from fibrillar muscles do not correspond to muscle contractions as they do in the Orthoptera, but they do provide an accurate picture of central nervous system output. In addition, the fibrillar muscles are unusual in that each comprises a single giant muscle fibre innervated by only a single fast motor axon. This makes it easy both to implant the electrodes into the muscles and to interpret the electrical events.

Using these techniques records have been made from most of the fibrillar and some of the smaller non-fibrillar muscles of *Drosophila melanogaster* during singing, and a summary of the 'motor tape' so obtained is illustrated in Figure 5.9. Using this information as a basis one can attempt to answer two questions: What is the underlying neuronal mechanism which produces this patterned motor output and, what are the mechanics of song production, that is, how do the muscles act so as to move the wings in the appropriate manner? With regard to the former one can again ask to what extent the pattern is endogenously controled and what role, if any, is played by sensory feedback.

Neuronal control

The neuronal mechanisms underling flight in several dipteran species including *Drosophila* have been extensively studied, and a model for the flight control system exists (Wyman 1973). The movements of the wings during song involve the same muscles as in flight and the two systems appear to be more similar than are flight and song in either crickets or grasshoppers. One might therefore expect to see similarities in the organization of the two behaviours in *Drosophila* and indeed this appears to be so.

By recording from two or more fibrillar muscles at the same time and examining the phase relationships of the muscle potentials, certain deductions can be made about the connectivity of the motor neurones which innervate the muscles. For example, during both flight and song the majority of the units within homonymous muscles fire with preferred phase relationships which strongly suggest that the motor neurones reciprocally inhibit one another. Close examination of the firing patterns indicates that this inhibition is monosynaptic and is not mediated by interneurones. This inhibitory relationship does not exist between the units of different muscles or between homologous muscles on the two sides of the body.

The rate of firing in all of the muscles fluctuates during the course of flight. However, although the rate for individual units can be quite different, the average changes synchronously in all of them. This indicates that they all receive a common excitatory input, possibly from a pacemaker circuit within the central nervous system. The idea of such a common input is strongly reinforced by the observation that during pulse song many muscle units fire with near synchrony. All of the fibrillar muscles fire with a fixed temporal relationship to the sound pulses, that is, about 18 milliseconds before each pulse. However, each of the units is active only before each second or third pulse. One

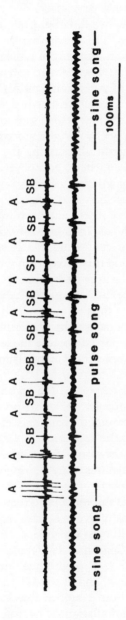

Figure 5.10. Example of a recording from an axillary muscle (A) and the sternobasalar muscle (SB) (upper trace) during *Drosophila melanogaster* courtship song (lower trace). (From Ewing 1979a).

presumes that the reciprocal inhibition that exists between homonymous units precludes any two such units firing during the same inter-pulse interval. In addition, one of the units, the first dorsoventral muscle, also fires at the beginning of sound pulses (Ewing 1977).

The non-fibrillar muscles also show two different preferred phases of firing with respect to the sound pulses. The various axillary muscles are active during the inter-pulse intervals and roughly in synchrony with the majority of the fibrillar muscle units. The basalar and sternobasalar muscles, by contrast, fire at the onset of every sound pulse (Figure 5.10). The observation that the second basalar muscle fires in synchrony with every wing beat not only during pulse song but also sine song and flight raises the possibility that the timing is set by peripheral feedback (Ewing 1979a). This was demonstrated by artificially driving the wing with a mechanotransducer at a variety of frequencies and observing that the potentials in the second basalar retained their synchrony with the forced wing movements. This feedback must be gated centrally, because the potentials were only observed once the insect had been induced to fly, and the frequency was then manipulated (Ewing 1979b). Heide (1983) has demonstrated that the first dorsoventral muscle is weakly phase locked to wing position during flight and this, along with the findings for song, suggests that feedback also influences its firing.

The sense organ involved in the feedback loop has been identified as pterale C, a structure protruding from the thoracic wall with which the wing collides at the mid-point of the downstroke. Recordings from the sensory nerve leading from this mechanoreceptor and the second basalar muscle show a clear reflex effect which is almost certainly monosynaptic (Miyan and Ewing 1984).

The model shown in Figure 5.11 incorporates the findings described above. All of the fibrillar and non-fibrillar motor neurones receive a common excitatory input from a presumed central pacemaker which produces a repetitive output at the pulse repetition rate for song. The timing of firing in fibrillar muscles is the result of this input, along with reciprocal inhibition between homonymous units. Peripheral feedback strongly affects the firing of dorso-ventral muscle 1 and the basalar and sternobasalar muscles. Confirmation of this model will require recording from the motor neurones in the central nervous system during singing. This will be a difficult task, but nevertheless possible, as the location of most of the cell bodies is known (Coggshall 1978) and intracellular recordings have been made from dorsal longitudinal muscle motor neurones in a static preparation (Koenig and Ikeda 1983).

Some aspects of the model have been confirmed and extended by results obtained from genetical manipulation. The role of sensory feedback is an example. It is probable that *Drosophila* can hear its own song via Johnston's organ, and this could be a source of feedback to set

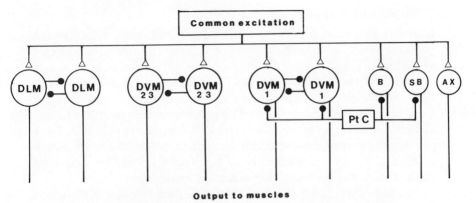

Figure 5.11. Model for the production of *Drosophila* pulse song. Excitatory input to the motor neurones from a common central source sets the song inter-pulse-interval. Reciprocal inhibition between homonymous units ensures that they do not fire synchronously. In addition, proprioceptive feedback to some motor neurones from pterale C (Pt C), and possibly other sense organs, influences the timing of firing of some of the muscle units. (Lettering as for Figure 5.9).

the timing of song pulses. Mutants which possess acoustically non-functional antennae nevertheless perform normally patterned songs, which eliminates this possibility (Burnet, Connolly and Dennis 1971). Feedback from a variety of mechanoreceptors on the wings and thorax such as pterale C could also serve the same purpose. In the *vestigial* mutant the wings are absent and, in some individuals, depending upon the penetration of the gene, the thoracic sclerites are 'locked' so as to preclude any perceptible movement. Nevertheless electrodes implanted into the flight muscles of such mutant flies show that, during courtship, the central nervous system produces the appropriate patterned motor output (Ewing 1979b). The recordings of Koenig and Ikeda mentioned above were only feasible because they made use of a temperature-sensitive mutant, *shibire*, in which the dorsal longitudinal muscle motor neurones fire spontaneously when the temperature rises above 28 °C. The neurones are otherwise silent except during flight and song.

The wing movements that occur during sine song are the same as those during flight, except for their low amplitude. Recordings from the fibrillar muscles during sine song show a much lower rate of firing compared with either flight or pulse song and, in addition, the more dorsal units of the dorsal longitudinal muscles remain entirely silent. These factors result in reduced tension developing in the muscles with the corresponding observed decrease in wing excursion. Here again we see a labile central organization in which the fly can switch rapidly between sine song and pulse song during its courtship.

The mechanism of wing vibration

It is more difficult to unravel the neuromuscular mechanisms underlying pulse song in *Drosophila* than those controlling stridulatory movements in crickets or grasshoppers. There is no clear concensus on the functions of many of the non-fibrillar muscles that are involved, and there is even some doubt concerning the role of some of the fibrillar muscles (Miyan and Ewing 1985a). Also, as the fibrillar muscles are myogenic, the timing of their contractions can not be ascertained from electromyograms. Thus the phase relationships between fibrillar muscle potentials and the sound pulses are merely a consequence of the neuronal circuitry described above and do not have any biomechanical significance.

It is likely that the fibrillar muscles are the prime movers of the wings during song. The timing of events and the known functions of the muscles suggest the following: Song starts with the wings folded in the rest position and one wing is progressively extended by contractions of the second basalar and sternobasalar muscles. At the same time, the axillary muscles which have not been individually identified, fire about twenty milliseconds before each sound pulse. This will have the effect of depressing the wing and stretch-activating the dorsoventral wing elevators by lifting up the dorsal scutum. The wing will consequently make the up-and-down movement responsible for a sound pulse (Miyan and Ewing 1985b). The sternobasalar muscle probably has a second function, that of damping wing movements and stopping the thorax from going into oscillation as in flight. This muscle, normally silent during both sine song and flight, during pulse song fires immediately after the onset of each sound pulse. Additional passive damping of wing movements is provided by those fibrillar muscle units which are not electrically excited during song. Again, mutant studies support this conclusion. The mutant allele, *stripe*, has the effect of preventing the development of some of the dorsal longitudinal units. Those flies with missing units have much louder songs, presumably due to the decreased degree of damping in the flight mechanism (Ewing, unpublished).

These results from *Drosophila* are very much in line with those from other singing insects. The temporal patterning of pulse song is controlled by an endogenous neuronal mechanism whose timing does not appear to be influenced by peripheral input. Some sensory input is normally required to initiate courtship, and it is known, for example, that the red eye of the female is one such stimulus (Willmund and Ewing 1982). The eye of the female also appears to be the stimulus which determines which wing should be extended during song; the wing nearest the female's head is always vibrated. Nevertheless such stimuli are not essential, and *Drosophila melanogaster* males will court in the dark where visual stimuli cannot be involved and chemical and/or tactile cues must be used. Once courtship has started, males will often continue singing even in the absence of a female.

6. The Functions of Arthropod Sound

In the widest sense the function of most arthropod sounds, except those which occur accidentally as a result of primarily non-acoustic activities, is communication, either inter- or intraspecific. However, it is not always easy to ascertain whether or not particular sounds are wholly or even partially concerned with communication. The high-frequency sounds produced by tsetse flies, *Glossina* species, during feeding and mating are usually considered to have some signal function in spite of the fact that no sense organ has been described which is capable of reception in the relevant frequency range. The observation that the sounds are associated with increases in body temperature of up to 12 °C above ambient and that the flies 'buzz' until their temperature reaches the optimum for take-off of between 30 and 32 °C suggests strongly that the buzzing is associated with an endothermic mechanism (Howe and Lehane 1986). This behaviour would be homologous with the warm-up thoracic vibrations that have been observed in many insects such as some moths and bumble bees (Kammer 1981). This does not, of course, preclude the possibility that the sounds are also used in communication in tsetse.

Similarly, bumble bees produce a characteristic buzzing which is different from the flight tone when they are gathering pollen from rose flowers, and it is thought that the vibrations that are produced loosen the pollen from the anthers and make it easier to collect. At the same time it is not impossible that the sounds are signals to other bees that the flower is already occupied (Heinrich 1979).

An observer may also attribute a display to the inappropriate sensory modality. Although a particular display may be perceived through more than one sensory channel it may be that the information through only one of these has any behavioural significance for the animal. Thus, for example, the wing vibrations of the *Drosophila* species, *D. pseudoobscura* and *D. persimilis* were recorded as sounds but initially the stimuli were thought to be visual and were discussed in relation to the flicker fusion frequencies of the insect eye (Waldron 1964). While the flies possibly do see the vibrating wing, the effective stimulus is almost certainly acoustic.

Nevertheless, one must be careful not to form a prejudgement based on what is known to be the case in related species. Flies of the genus *Drosophila* provide a good example of this, as the great majority of species whose courtship behaviour has been examined perform some form of wing display. In most of these the relevant stimuli produced are undoubtedly acoustic, but in some, such as *D. suzukii* and *D. biarmipes*, the displays, although producing sound, are essentially visual. A clue to this is the presence of a black shade on the wings of some individuals of these species. *D. biarmipes* is dimorphic for this trait, and those individuals with the dark markings are sexually more successful than are those without, which reinforces the argument for the visual nature of the stimulus (Singh and Chatterjee 1987).

Other examples of possible ambiguity are found in the displays of fiddler crabs and the courtships of some grasshoppers. The exaggerated size and the colouration of the chelae in the former, along with their complex patterns of movement, provide strong circumstantial evidence for visual displays but, as described in Chapter 10, they also produce behaviourally relevant vibrational signals which might easily be overlooked. Most grasshopper species perform displays which involve movements of the hind legs. In some, these are clearly acoustic, while in others the knees are brightly coloured and are used in a visual display, and in yet another species the knee waving is accompanied by vibrational signals which are produced when the animals are out of visual contact (Riede 1987). Some acridids also perform crepitating flights which could provide either acoustic or visual or both stimuli at the same time.

In addition to the possible ambiguities which can arise because sounds may result from essentially non-communicatory behaviour, or from the irrelevant by-product of a non-acoustic signal, many sounds may have more than one function. For example, while many insects use sounds during courtship they may also, like some cicadas and the cockroach, *Gromphadorhina*, produce very similar sounds when handled. In this context they are forms of startle display designed to deter potential predators.

It is important to keep the above provisos in mind when considering the possible functions of acoustic and vibrational signals. The evidence for function in the great majority of cases is based upon their behavioural context and no experimental verification exists. While this is often adequate, there are many examples where the precise functions of acoustic behaviour are by no means clear.

I shall deal with the functions of acoustic and vibrational communication under three headings: interspecific, intraspecific within social animals and intraspecific in non-social animals. The last category is mainly concerned with sexual, territorial or aggressive behaviour.

INTERSPECIFIC COMMUNICATION

Interspecific acoustic interactions are almost all defensive in nature and are of three main types. The behaviour may be aposematic in function and advertise the unpalatability of the prey; it may be mimetic or it may be designed to startle. It is unfortunate that there is little experimental verification for these functions in the great majority of examples that are quoted in the literature. Much of the observation is anecdotal in nature and the evidence indirect or circumstantial.

Startle responses

Many arthropods produce sounds when disturbed or handled by means of a variety of mechanisms. The majority of these are probably responses analogous to the visual startle displays of some insects. Well-known examples are those of butterflies and moths which are normally cryptic at rest and open their wings to reveal bright patterns, often in the form of eyespots. One has only to observe the reaction of a naïve person who picks up a specimen of the cockroach, *Gromphadorhina portentosa*, which responds by producing a series of loud hisses, to see the efficacy of the mechanism.

Cicadas, some tettigoniids, beetles, bugs and cockroaches among others produce sounds, usually stridulatory, on being handled. In some cases these sounds do not appear to be produced in other contexts, suggesting that their primary, and perhaps sole function is defensive. In those species which use sound for intraspecific communication as well, the defensive sounds tend to be more irregular and unpatterned, which might enhance their startle value and at the same time reflects the fact that patterned sound is important for intra-specific communication. In some cases, however, defensive sounds are patterned, as in the New Zealand wetas, and here one finds that there is very little difference between the sounds produced by the different species (Field 1978). In such cases the repetitive nature of the sound pulses may merely reflect the underlying neuronal mechanisms driving the muscles, as occurs during flight or locomotion, and it may not have any communicative significance.

For the reasons discussed in Chapter 2, the songs of many insects have rather narrow frequency spectra tuned to the properties of specific receptor systems. By contrast, one might expect warning or startle sounds to have a wide frequency spread, so that they would be heard by a range of different potential predators, which would, in the majority of instances, be vertebrates. There are very few examples recorded of a warning sound deterring an invertebrate predator. Where the spectra of such signals have been examined in, for example, the tenebrioid beetle, *Adelium pustulosum*, (Eisner, Aneshansley, Eisner, Rutowski, Chong and Meinwald 1974) and the bug, *Coranus subapterus* (Haskell 1957) this appears to hold (Figure 6.1).

An exception where the warning signal is tailored for one specific type of predator appears to be the case in the peacock butterfly, *Inachis io*. It produces a series of high-frequency clicks in response to substrate

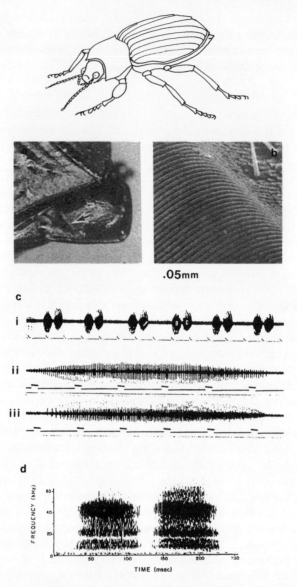

Figure 6.1. Sound production in the beetle, *Adelium pustulosum*. a, Abdominal tip showing the extent of the stridulatory movement. Files (arrow) at the sides of the abdomen scrape against the edge of the elytra; b, Scanning electronmicrograph of the stridulatory ridge; C, oscillograms of the chirps. i, A chirp sequence: the doublet chirps are due to outward (B) and then inward (C) movements of the abdomen (time marker, 200 ms); ii, iii, Expanded records of B and C (time marker, 20 ms); d, Sound spectrogram of a bisyllable chirp, demonstrating the wide range of frequencies. (From Eisner *et al.* 1974).

vibration. In the hibernaculae which the overwintering butterflies often share with bats, the sounds are triggered by the approach of a bat, and the immediate response of the bat is to start back and retreat. The frequency of the butterfly sounds lies between 30 and 60 kHz and closely matches the bat's auditory receptor mechanism related to echolocation. It is worth noting that just as birds become habituated to the eyespots on the wings of the peacock butterfly, so too do bats fail to respond to the ultrasonic clicks which may ultimately attract rather than repel (Mohl and Miller 1976).

Another supposed example of a defensive signal being tailored to a specific type of predator has been proposed for the warning stridulation of the cockroach, *Henschoutedenia epilamproides*. This insect stridulates by means of a scraper and file on the posterior edge of the pronotum and the dorsal surface of the metathorax, and does so in response to handling or to a shadow. The chirps are strongly amplitude-modulated with a carrier frequency of around 5 kHz and with many strong side-bands (Guthrie 1966). These chirps are very similar in structure to the squeaks of rats, which are known to be important predators of cockroaches. Unlike many cockroaches, *H. epilamproides* is not distasteful and its stridulations might initially confuse a rodent predator due to the similarity in their vocalizations. However, the effect of cockroach stridulation on the behaviour of the rat does not appear to have been established.

There have been a few studies of the effects of defensive sounds of arthropods on potential predators. A comparison of the ways in which the grasshopper mouse, *Onychomys torridus*, treated noisy and silenced cicadas showed that they were less efficient in dealing with the former. They took longer to overcome them and made more handling errors, which allowed some cicadas to escape (Smith and Langley 1978). Stridulation by passalid beetles conferred a similar advantage when they were attacked by crows. As these beetles are subsocial it might be expected that their defensive stridulations would act to warn others in the colony, but no evidence for this was found (Buchler, Wright and Brown 1981). Less clear-cut results were obtained by Alexander (1958) with scorpions. Their stridulations were shown to be perceived by a wide variety of vertebrates, but it was not obvious what advantage was conferred on the scorpions by the production of sound, as the potential predators that were tested did not appear to be repelled by them.

Less is known of the responses of invertebrate predators to sound and vibration. Lycosid spiders tested with a variety of stridulating beetles, *Tropisternus* spp. and *Omophron labiatus*, and mutillid wasps, *Dasymutilla* spp., were more successful if the prey were first silenced. The spiders could also be induced to attack a wire probe and they were less persistent if the probe was vibrated so as to mimic a stridulating insect (Masters 1979).

Not all defensive sounds are produced by adults and pupal sound production is widespread among Coleoptera and some families of Lepidoptera. Many beetle pupae possess defensive structures called gin-traps which consist of tubercles or teeth on the mating surfaces of adjacent abdominal segments. Small insects or mites and also the appendages of larger arthropods become caught and perhaps crushed in these structures as the abdomen moves in response to tactile stimulation (Hinton 1955). The structure of the gin-trap is preadapted as a stridulatory device and in some instances, as in the dynastid beetle, *Oryctes rhinoceros*, it serves both functions. However, it is in the Lepidoptera that pupal sound production is most common, and the pupae of many species of hawk-moth (Sphingidae) have modified gin-traps whose primary function appears to be stridulatory.

A different kind of pupal stridulation is found among noctuid moths of the subfamilies Nyctiolinae (= Sarrothripinae) and Westermanninae (= Careinae, = Chloephorinae). In the majority of stridulating species the dorsal surfaces of the head, thorax and part of the abdomen are covered with tubercles and these scrape on the inside of the wall of the cocoon to produce sound. An added sophistication in some of the Westermanninae is that the cocoons have a series of hard, longitudinal ridges on their inner surface which act as a file across which the modified 10th abdominal segment is drawn (Hinton 1948).

Percussion appears to be used by pupae of some skipper (Hesperidae) and blue butterflies (Lycaenidae) which tap parts of their bodies against the substrate on which they have pupated, thus making audible rattling or knocking sounds. There appears to be little direct evidence for the protective function of this behaviour, and some of the sounds that are produced could be incidental. However, normally they are produced only in response to disturbance and it is difficult to see what other function they might serve.

Acoustic mimicry

It is not at all clear to what extent acoustic mimicry has evolved as a specific defensive strategy as opposed to being merely fortuitous. Many hover-flies (Syrphidae) which are visual mimics of a variety of Hymenoptera such as wasps, bumble and honey bees, have flight tones which are similar to their models. This is also true of various parasites such as the bee flies (Bombylidae) which are parasitic upon solitary bees, and the cuckoo bees (*Psythyrus* species) which parasitize true bumble bees (*Bombus* species). In these cases there is no evidence showing to what extent the similarity between model and mimic protects the latter from its model rather than from predators. The buzzing produced by the mimics when captured certainly increases one's apprehension until the insect is correctly identified, and would probably act as an additional

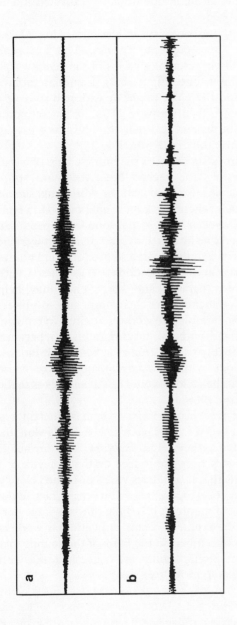

Figure 6.2. Oscillograms of the flight tones of (a), the Hornet (*Vespa crabo*) and (b) its visual mimic, the Hornet Clearwing Moth (*Sesia apiformis*). (From Rothschild 1985).

deterrent to a potential predator. However, as both Diptera and Hymenoptera have asynchronous flight muscles, and as models and mimics are the same size, a similar flight tone would be expected and no evolved mimicry need be invoked.

A better case for acoustic mimicry exists in the Clearwing Moths (Sesiidae) which also mimic hymenopteran species. The European Clearwing, *Sesia apiformis*, closely resembles wasps of the genus *Vespa* and also has a very similar flight tone (Figure 6.2). Lepidoptera have synchronous flight muscles and therefore might be expected to have a much lower wing beat frequency than wasps ($c.\,110\,\mathrm{s}^{-1}$). These moths are themselves distasteful to birds, and this may therefore be a case of Müllerian rather than Batesian mimicry (Rothschild 1985).

A rather special relationship exists between the bumble bees (*Bombus* species) and their nest parasites, the cuckoo bees (*Psythyrus* spp.). *Bombus* females are known to make use of acoustic or vibrational signals within their nests, although their function is largely unknown. However, it has been shown that when a queen is removed, sounds are produced which are associated with aggressive interactions, and dominant workers gain the right to lay eggs. *Psythyrus* individuals also produce loud buzzes when occupying *Bombus* nests, and these sounds are associated with aggressive interactions between parasite and host. They may help *Psythyrus* females to assert dominance over their hosts and so protect their eggs and brood which the bumble bee workers attempt to remove. The buzzes are fairly wide-band with the maximum energy between $700\,\mathrm{Hz}$ and $2\,\mathrm{kHz}$ and are probably transmitted as substrate vibrations. Artificially produced vibrations within this frequency range cause bumble bees to 'freeze', and this behaviour would be advantageous to the nest parasite (Fisher and Weary 1988).

Another intriguing example of acoustic mimicry which is often quoted concerns the association between the Death's Head Hawkmoth, *Acherontia atropos*, and the honey bee, *Apis mellifera*. The former is known to enter the hives of honey bees and to feed on the honey, and it has been noted that the sounds that it produces when disturbed closely resemble the 'piping' of queen bees (Figure 2.8). However attractive the idea that this constitutes a case of mimicry, it is difficult to envisage how it would be of advantage to the moth. It seems unlikely that workers would mistake the moth for a queen bee on the basis of the sounds that it produces, particularly as the piping sounds are concerned, not with recognition, but with the preparation for swarming.

Aposematic functions of sound

Many moths within the families Arctiidae and Ctenuchidae are avoided by vertebrate and some invertebrate predators. It is clear from the behaviour of birds and monkeys induced to eat them that they are extremely distasteful. Their larvae feed mainly on toxic plants and sequester a variety of poisonous and unpleasant-tasting chemicals such as cardiac glucosides and pyrrolizidine alkaloids. The adults of many species are brightly coloured, predominantly red with yellow or white. In addition some arctiids, for example *Utetheisa* and *Rhodogastria* species exude a noxious foam from cervical glands when disturbed. These, and other defensive devices, are described by Blest (1964) and it is clear that the colouration and displays have an aposematic function. As described in Chapter 2, these moths are capable of producing trains of high-frequency clicks by means of a thoracic tymbal mechanism. These sounds may be similarly aposematic and would be effective against nocturnal predators such as bats.

It has been demonstrated clearly that bats do indeed avoid these moths when they produce trains of clicks. Furthermore, Dunning (1968) also showed that the moths produced the sounds in response to the cries of echolocating bats. Bats trained to catch meal-worms thrown into the air avoided them if the meal-worms were presented along with recorded moth clicks played through a loudspeaker (Dunning and Roeder 1965). As the moths do not appear to respond to conspecific sounds, the evidence points to the primarily defensive nature of the ultrasonic sounds. The reason for this defensive advantage is, however, not clear and alternative explanations can be put forward.

One hypothesis is that, as the clicks are similar in structure to the echolocating cries of some bats, they act as jamming devices. However, many experiments, from those of Spalanzani at the end of the eighteenth century, up to modern times, have demonstrated the resistance of bats to acoustic jamming, and this idea remains speculative (Sales and Pye 1974, Miller 1983).

The more generally accepted idea is the one outlined above, that the moth sounds are indeed aposematic and that bats learn to associate the clicks with the distasteful characteristics of the moths. Both Blest (1964) and Dunning (1968) noted that not all of the different species were equally repellent, and that the less distasteful the moth was, the more easily were the defensive sounds elicited, either by handling or in response to bat cries. It appears that the effectiveness of ultrasonic sound production in these moths has become enhanced by a degree of Müll-erian mimicry, while at the same time the better tasting species are Batesian mimics of the more noxious ones. An analogous situation has arisen more than once in complexes of mimetic butterfly species (Turner 1984).

It is slightly unexpected that the characteristics of the sounds in different species are very variable, both with regard to pulse repetition rate and carrier frequency, with the latter, for example, ranging from 19 to 85 kHz. One might expect that selection pressure would have acted to

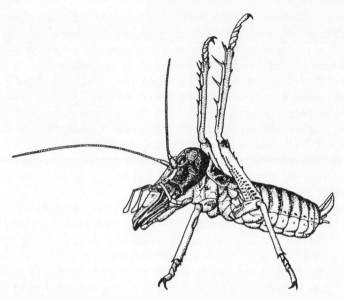

Figure 6.3. Threat posture of the Weta, *Hemideina thoracica*. The animal stridulates as the legs are kicked down, directing the tibial spines towards the source of disturbance. (From Field 1978).

make them converge. However, if the stimulus were sufficiently aversive, a bat would generalize from one to a wide range of ultrasonic signals. In support of this, bats do not appear to discriminate the calls of different moth species, although Dunning (1968) gives the example of one individual which, after a long period of being used in an experimental situation, learned to distinguish the calls of a patatable species, *Pyrrharctia*, from those of distasteful ones. This bat was able to make the distinction when *Pyrrharctia* responded, not to its echolocating calls, but to those of other bats. This would allow it more time to evaluate the information, and the development of this ability was probably an artefact of the experimental set-up. It does show nonetheless that bats are potentially able to discriminate different species of moth on the basis of the sounds they produce.

In the majority of cases it would be unwise to ascribe a single mechanism to any of the anti-predator strategies. Startle, warning and aposematic functions may all be combined and modalities other than acoustic may also be involved. Many warning sounds are accompanied by visual displays, while the release of unpleasant chemical substances is also common, as found in some arctiid moths and the beetle, *A. pustulosum*. The Weta produces a defensive kick when attacked, which, considering the size of the insect and the spines on its hind legs, would be likely to be most effective. The kicking movement also actuates the stridulatory mechanism and the sound would reinforce the effect of the kick. Interestingly, the same structures are involved in intraspecific stridulation, but in this context it is the abdomen which is moved across the femora (Figure 6.3)(McVean 1986).

Acoustic interception

There is one form of interspecific communication not concerned with defence which merits consideration: the interception of acoustic signals by predators or parasites. This is a considerable potential hazard for many insects which produce loud calling songs for long periods of time. These songs are designed to advertise the presence of the signaller and possess characteristics which allow them to be easily located by conspecifics. Selection will have acted to produce songs which are maximally effective in intraspecific communication while, at the same time, being safe from unwanted interception. It is difficult to disentangle which song features result from selection for these two partially incompatible aims, but more than one strategy appears to have evolved.

Some calling insects have been shown to modify their behaviour in response to the cries of hunting bats. The tettigoniid, *Insara covilleae*, and the Wax Moth, *Achroia grisella*, which have calling songs in the ultrasonic range, fall silent when echolocating bats fly past. Both of these species call from exposed situations and may well be liable to predation by bats although this has not been observed to happen (Spangler 1984).

Certain sounds are more difficult to locate than others. Narrow-band, high-frequency sounds do not permit phase differences to be used for direction finding. Sounds lacking transients or amplitude modulation are also difficult to locate. The trilling songs of some cicadas and tettigoniids have both of these features, and anyone who has attempted to track down such insects using their songs will testify to the difficulty of the task. At the other extreme, very short bursts of song widely spaced in time will also provide minimum information for a predator. Yet another strategy used by some fireflies (Lampyridae), that of decreasing stimulus intensity as the communicants approach one another, does not seem to be employed by singing insects. This is possibly because for many kinds of acoustic mechanism this would not be feasible due to the mechanical constraints of the systems. However, some species change from a long-distance, acoustic signal to a close-range vibrational one when the signaller and recipient are close enough to communicate by this means. This will have the effect of decreasing the possibility of intervention by either a predator or a conspecific rival.

One example of a parasite utilising the calling song of an insect in order to find its host is the larviparous tachinid fly, *Euphasiopteryx ochracia*, which parasitizes various *Gryllus* species. The flies can be induced to approach and larviposit on a loudspeaker from which the calling song of a cricket is being broadcast (Cade 1975). There is a difference between the patterns of calling in cricket species which are liable to parasitism and those which are not. *G. intiger* is parasitized and calls for a much shorter time and over a more variable period than the closely related *G. pennsylvanicus* and *G. veletis* whose songs do not attract the tachinid flies. This possibly results from selective pressures due to parasitism (Cade and Wyatt 1984).

Tachinid flies are also attracted to other calling Orthoptera, including mole crickets (*Gryllotalpa*) and bush crickets (Burk 1982, Mangold 1978). A similar case has been reported in cicadas where a sarcophagid fly, *Colcondamyia auditrix*, finds its host by its calling song (Soper, Shewell and Tyrrell 1976).

There are surprisingly few examples reported of predators using songs to locate prey. As most insects stop calling once they perceive vibrations, it is difficult for a terrestrial vertebrate predator to approach a singing insect undetected. Nevertheless, cats, toads and herons have all been described as being attracted to calling Orthoptera (Walker 1964a, 1979, Bell 1979). I have observed my cat locating grasshoppers by their songs and attempting to catch them in the garden. As grasshoppers form a substantial portion of the diet of feral domestic cats, song clearly has drawbacks.

Aerial predators would be less disadvantaged in locating prey by song although, as described above, bats can be detected by their echolocating cries. Because of their high carrier frequencies, the calling songs of some tettigoniids would be detected by bats, which might therefore be expected to locate their prey by this method, and workers have been aware of this possibility for some time. However, the first demonstration of acoustic clues being used in this way by bats is in species which predate forest anurans. Tuttle and Ryan (1981) demonstrated that neotropical bats responded to the calls of edible frogs and, further, that they were able to discriminate the calls of edible and poisonous species. They were also able to determine the size of the species from its call, and did not attempt to predate frogs that were too large. The bats were therefore capable of making complex discriminations on the basis of the species-specific anuran calling songs.

Subsequently, in 1987, Belwood and Morris at the same study area of Barra Colorado in Panama showed that bats also predated tettigoniids (katydids) using acoustic cues. They were able to mist net four bat species, including *Trachops cirrhosus*, used by Tuttle and Ryan in their study, by using male singing tettigoniids as bait. Silent females, although highly predated by the bats, did not attract any bats into the mist nets. Such predation pressure would be expected to modify the calling behaviour of tettigoniids (and anurans). A comparison of the tettigoniid species inhabiting forest clearings, which are not predated by bats, with those which call from foliage within the forests and are predated, showed

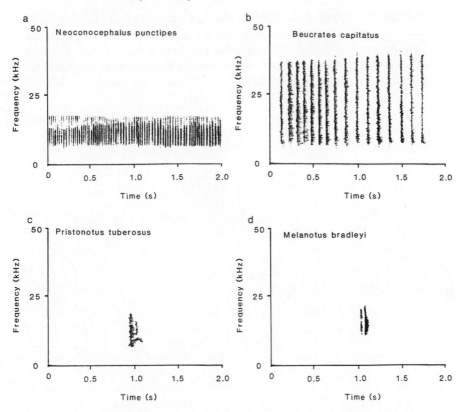

Figure 6.4. Audiospectrograms (diagrammatic) of two forest-dwelling bush-crickets (a,b) compared with those of two clearing species (c,d). The songs of the latter are shorter and tend to have a narrower frequency range. (After Belwood and Morris 1987).

that songs of the latter contained much less sound per unit time, that is, they had a much lower duty cycle than the songs of forest-clearing species. In addition, they contained a narrower range of frequencies (Figure 6.4). These two factors would make singing forest-dwelling species less liable to detection and thus predation by bats. This suggests that predation can be an important selective force, shaping the calling songs of insects, a conclusion also reached with respect to anuran species.

Sakaluk and Belwood (1984) have shown that the Mediterranean House Gecko, *Hemidactylus tursicus*, also orients to the calling songs of the cricket, *Grylloides supplicans*. Males of this cricket call from within burrows and are not themselves predated. However, the geckos intercept and catch females which are attracted to male calling song. *H. tursicus* is an introduced species in Florida and it is not known if it uses this strategy in its native habitat or what its prey is there. In the Barra Colorado study it was observed that female tettigoniids constituted a large proportion of the bats' diet even although they do not themselves call. Clearly phototactic behaviour can make females more conspicuous and therefore liable to predation.

7. Intraspecific Acoustic Communication: calling songs

A fundamental need for the individuals of any animal species is for the sexes to come together to mate. There will be strong selective pressure for this to occur in the most economical fashion, that is, without the participants being exposed to undue risk from predation or using excessive energy. It is also necessary for them to make correct decisions concerning species identity and physiological suitability. In the arthropods these requirements are achieved through a wide variety of behavioural and ecological mechanisms. Chemical cues, either pheromonal or in the form of food plant odours are arguably the most widespread means of bringing the sexes together (Carde and Baker 1984). However, acoustic and vibrational cues are also extensively used, either alone or in conjunction with chemical and visual stimuli.

While acoustic communication is used in diverse behavioural contexts, it is the 'calling songs' of some crickets, bush-crickets and cicadas which, having evolved to advertise the presence of the insect, are most often heard. Although the primary function of many of these songs is to attract potential mates, they may be additionally concerned in other behaviours such as spacing and territory holding or, alternatively, in forming chorusing aggregations.

In almost all cases songs are produced by males which call from some fixed position, the nature of which depends upon the species: mole-crickets call from their burrows, bush-crickets usually from low herbage and cicadas from trees. As described in Chapter 6, calling makes insects liable to predation and parasitism. That males rather than females sing accords with one of the tenets of population genetics that the sex which bears the smaller parental investment will incur the greatest risk (Trivers 1972). This is an attractive hypothesis which goes a long way to account for the sexual dimorphism or, conversely, the lack of it, in many bird and mammal species. It is more difficult to obtain satisfactory supportive evidence in arthropods.

In the majority of insects it is the female which, as the egg producer, and sometimes guardian of the progeny, has the larger parental investment. However, the evidence presented in Chapter 6 on predation of crickets by geckos and bats shows that the former predated not the calling males but the responding females. Bats preyed equally on calling male bush-crickets and on the females which were made more conspicuous by moving towards the males. Each of these examples deals only with the effects of a single type of predator, and the overall pattern of predation needs to be known before the true costs and benefits of singing and phonotaxis for the two sexes can be evaluated.

Some support for a negative correlation between degree of parental investment and acoustic advertisement is found in the Mormon Cricket, *Anabrus simplex*, which produces a large spermatophore up to 27 per cent of the total body weight. In addition to the sperm this contains a considerable amount of protein which the female makes use of in producing her eggs. The male's investment is therefore considerable, and could be even larger than that of the female, in which case one might expect a reversal of sexual roles. The males of this species, unlike other crickets, do not call for protracted periods but for a maximum of only two minutes, which suffices to attract females. Males then select among the females for larger and therefore more fecund individuals (Gwynne 1982).

Assessment of calling song

Advertisement displays, whether acoustic or vibrational, or indeed using any other sensory modality, raise two distinct categories of question. The first concerns the specific function of the behaviour. Is it designed to attract conspecifics, males, females or both? Does it have an additional territorial or spacing function? The second relates to the mechanisms whereby these functions are achieved. Which acoustic parameters are important in eliciting phonotaxis? What cues are used in species identification or mate choice or in deciding social status?

Different experimental approaches have been used to find answers to such questions. While the functions of some acoustic behaviour can be deduced from ethological observations and simple behavioural experiments, the analysis of mechanisms normally requires more intrusive techniques. The phonotactic response of female orthopterans, particularly crickets, has been extensively used as a tool in such studies.

Brief descriptions of the methods used by different workers to measure phonotactic responses are appropriate because very disparate techniques have been used. In addition, conflicting results have been obtained which probably result from methodological differences. At the simplest level, observations of females attracted to calling males or to artificially broadcast songs in the field can provide very useful information, particularly in regions where more than one species is active. It could be argued that, no matter what results are obtained in the laboratory, they should always be checked against field observations of this sort.

The majority of laboratory-based experiments have made use of an open field arena of some type or a Y-maze where the phonotactic responses of females to caged calling males or to recorded or synthesized

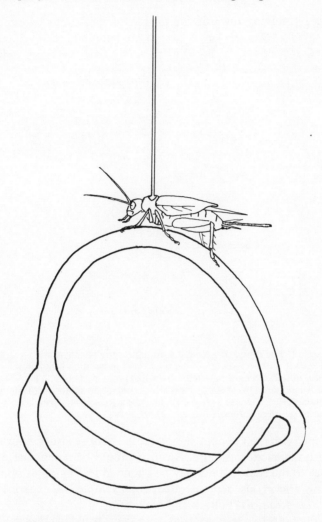

Figure 7.1. A tethered female cricket walking on an expanded polystyrene (styrifoam) Y-maize. Her phonotactic preferences can be assessed by analysing the choices she makes in response to songs played to her two sides.

song can be assessed. These experiments can be difficult to interpret because of the choice of appropriate response criteria. Often females are required to make a differential response and choose between two sound sources. The sounds may be presented either simultaneously or alternately, with the two methods sometimes providing rather different results. In many Orthoptera and in cicadas phonotaxis occurs during flight, and a more appropriate method of testing is by using a suspended insect in tethered flight. Crickets and other insects make steering

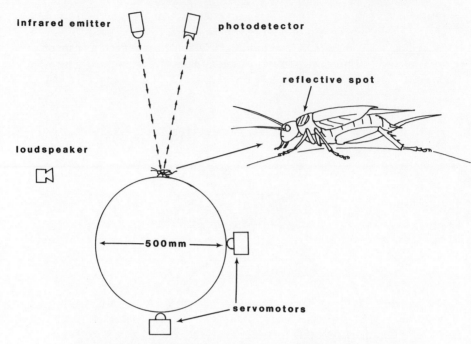

Figure 7.2. The spherical treadmill used to measure phonotaxis in crickets. The cricket walks on the top of the sphere and its position is monitored with a beam of infrared light reflected from a spot on its pronotum. The reflected light is monitored by photodetectors which control the motors adjusting the position of the sphere so as to ensure that the cricket always remains at a constant position. (From Huber and Thorson 1983).

movements with their abdomens and these can be recorded to provide an assay of phonotactic response. Again it should be noted that different responses to the same sound may be obtained in a flying as opposed to a walking cricket, as demonstrated, for example, in experiments on *Teleogryllus oceanicus* (Pollack and Hoy 1979).

The phonotactic responses of tethered, walking insects can be measured also by means of an ingenious device, the 'Y-maze globe' (Figure 7.1). This was initially designed by Hassenstein (1959) to investigate phototactic behaviour in the beetle, *Chlorophanus*. It consists of a spherical maze constructed from straws or styrifoam which the insect grasps with its tarsi and rotates as it walks. As it does so, the insect is presented with a succession of left/right choices which can be scored. This has the advantages that the behaviour can be readily quantified and that both the position and the intensity of the acoustic stimuli can be precisely controlled. In a traditional Y-maze both of these factors may change as the female approaches a sound source.

A sophisticated version of the Y-maze globe, the 'Kramer treadmill', has been developed by Weber and his co-workers (Weber, Thorson and

Huber 1981) to study phonotaxis in crickets. An untethered female cricket walks on the top of a large sphere which rests on, and is itself moved by servo-motors. A spot on the cricket's thorax reflects a beam of infrared light. This is monitored by photodetectors whose response is used to control the drive on the servo-motors. This is arranged so that the cricket is kept at the top of the sphere regardless of the actual direction in which she walks. The corrective signals are stored and provide a detailed record of both of the intended direction and speed of the cricket's movement (Figure 7.2).

FUNCTIONS OF CALLING SONGS

The evidence for the role of male calling song in attracting conspecific females is overwhelming in several insect groups. There are two ways in which the species specificity of these songs has been tested. Phonotactic experiments comprise one such method. However, in many insects the communication is two-way, that is, the female answers the calling song of the male with one of her own. The female acoustic response can be used as a powerful behavioural assay for recognition of male song.

The first clear demonstration of species-specific phonotaxis was probably in the crickets, *Oecanthus* species, from the eastern United States (Walker 1957). Walker recorded the songs of nine different species and showed that they had distinct patterns of trills or chirps with characteristic pulse and chirp repetition rates. He tested, in a linear arena, two pairs of sympatric species each living in similar habitats, one pair from trees and the other found in low undergrowth. A female was placed in the centre of a rectangular area and the song of one of the species was broadcast from one end, the other end providing no acoustic stimulus, and the movement of the female noted. For the species from the same habitat, he showed that females were attracted only to the song of conspecific males but not to those of heterospecifics. He also tested females with synthetic songs whose parameters he could alter, and demonstrated that it was the temporal patterning and not spectral composition of the songs which was important in determining female choice.

Other similar examples are provided by Hill, Loftus-Hills and Gartside (1972) who were able to show that females of the Australian Field Crickets, *Teleogryllus commodus*, and *T. oceanicus* each chose the songs of conspecific males as did females of the European Crickets, *Gryllus campestris*, and *G. bimaculatus* (Shuvalov and Popov 1973). Each of three very similar, sympatric ground crickets, *Allonemobius* species, also preferred homospecific song if given the choice of its own and that of another species. If, however, they were presented with only one song type they were attracted to it regardless of which species it belonged to (Paul 1976). This is different from what was found in *Oecanthus* species which apparently ignored heterospecific song.

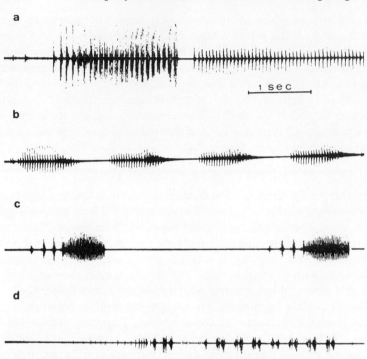

Figure 7.3. Calling songs of gall midges, *Lipara* species. The calls of males are distinct. Those of the females differ only in their durations. a, *L. lucens* male song followed by part of the female's answer; b, c, d, male songs. b, *L. similis*; c, *L. pullitarsis*; d, *L. rufitarsis*. (From Chvala *et al.* 1974).

Rather few experiments have been carried out under field conditions designed specifically to demonstrate the selectivity of calling song responses. Ulagaraj and Walker (1973) showed that the Mole-crickets, *Scapteriscus acletus*, and *S. vicinum* were preferentially attracted to loudspeakers broadcasting both natural and synthesized songs of their own species. It is interesting that the sound-baited traps caught a small proportion not only of the heterospecific mole-cricket species but also three other species of gryllid, demonstrating that the discrimination based on song is not perfect.

Female acoustic responses have been relatively little used to assess song parameters. A classical ethological analysis of two grasshopper species, *Chorthippus biguttulus* and *C. brunneus* by Perdeck (1957) showed that it was male song alone of all of the possible types of stimuli which elicited female stridulation, and he concluded therefore that calling song was the primary mechanism for pre-mating isolation in these species. Other insects in which species-specific, male–female acoustic interactions have been noted include water bugs of the genera *Cenocorixa* (Jansson 1973), *Palmacorixa* and *Sigara* (Jansson 1976) and stoneflies, *Isoperla* species (Szczytko and Stewart 1979). The latter communicate by substrate vibrations and some species show three-way communication with males responding to the female's answer with yet another song pattern. Gall midges of the genus *Lipara* also have vibrational signals, and males fly between *Phragmites* stems producing pulsed vibrations which are transmitted along the reed stems to be answered by receptive females. It is notable that while the songs of males of different species are distinct, those of the females are similar to one another (Figure 7.3)(Chvala, Doskpcil, Mook and Pokorny 1974). This is to be expected on the basis of the female's greater parental investment; the correctness of her discrimination is the more important. This feature is seen in many species of insect which perform duets.

In all of the examples detailed above some degree of selective species-specific response is involved, and calling songs are one of the factors responsible for bringing conspecifics together. It follows that some type of recognition system must exist within the central nervous system, either a filter mechanism or a song 'template' as has been proposed in birds (Konishi 1965). The search for this recognition system described in Chapters 3 and 4 has attracted much interest. However, there are several features which will have to be incorporated into any eventual model, such as a considerable degree of flexibility and susceptibility to extrinsic modification. The recognition of conspecific song is seldom perfect and, while this may merely reflect imperfections of the underlying mechanisms, this is not necessarily so.

Where species are not sympatric, even when their songs are very dissimilar, discrimination based on song is often poor. This is seen, for example, in *Epippiger* species (Dumortier 1963) and in bush-crickets (= katydids) of the genus *Conocephalus* studied by Morris and Fullard (1983) and Gwynne and Morris (1986). These workers showed that while the females of *C. nigropleurum* discriminated accurately against songs of the sympatric species, *C. brevipennis*, they responded positively towards those of allopatric species and even to a random pattern of noise so long as the main energy peak was in the appropriate frequency range. It makes sense in evolutionary terms if females respond, either negatively or positively, only to songs of sympatric species. The selective pressures which could bring this about are also clear, but it is nevertheless difficult to envisage a recognition system with these properties.

A further difficulty is that response to song can change dramatically with prior experience of song and, while the initiation of phonotaxis may require a precisely defined acoustic signal, the response, once initiated, may continue towards a very inadequate or imperfect song model. This has been demonstrated from a wide range of insects such as grasshoppers (Skovmand and Pederson 1978), crickets (Zaretsky 1972) and stoneflies (Zeigler and Stewart 1986). This property of the recognition system will eventually have to be taken into account when explaining acoustic signal processing.

PHONOTACTIC STRATEGIES

There are two general patterns of phonotactic behaviour. In the simplest, found in cicadas, most gryllids and tettigoniids, the male sings loudly and fairly continuously and the female approaches him. A more complex pattern occurs when the female replies to the song of the male and a duet follows during which one or both of the sexes may move towards one another. There are many variations to be found in the latter category, both in the way in which the songs interact and in the patterns of movement. Duetting is the rule in the Acrididae and in some Hemiptera, Plecoptera and Tettigoniidae.

In the gomphocerine grasshoppers, whose behaviour and neurobiology have been so extensively studied, males do not call from a stationary position but move around stridulating intermittently until answered by a female. The male and female both sing and approach one another. This allows the male to cover a large area. Although his movement presumably makes him liable to predation, this increased risk will be offset by the decreased risk of attracting a predator or parasite by his song. An additional advantage is in decreased energy expenditure devoted to singing. The songs of stationary, calling species are always much louder than those of ambulatory ones. Many static calling insects, such as crickets and cicadas, are 'cuckolded' by males of their own species, satellite males, which do not themselves call but intercept approaching females and mate with them. A moving male is less liable to be tricked in this way.

Similar duetting behaviour has been observed in stoneflies, gallmidges and land and water bugs as mentioned above. In the Leafhopper, *Amrasca devastans*, which communicates by means of substrate vibrations, males call and are answered by conspecific females. The effects of this exchange of signals are to increase levels of locomotion in males which move towards the females while at the same time inhibiting female movement (Saxena and Kumar 1984). In the Water-bug, *Corixa dentipes*, only those copulations in which duets are performed are successful. When males manage to mount non-responding females, no sperm is transferred (Theiss 1983). The female reply to male calling song is not therefore merely a species identification signal but also indicates readiness to mate. Males are much more promiscuous than females and attempt to mate with heterospecific females as well as their own.

Interesting variations in phonotactic strategies are to be found in bush-crickets (katydids) of the subfamily Phanopterinae. The behaviour of some North American species has been described by Spooner (1968a) who subjectively describes their vocalizations as 'ticks', which are very

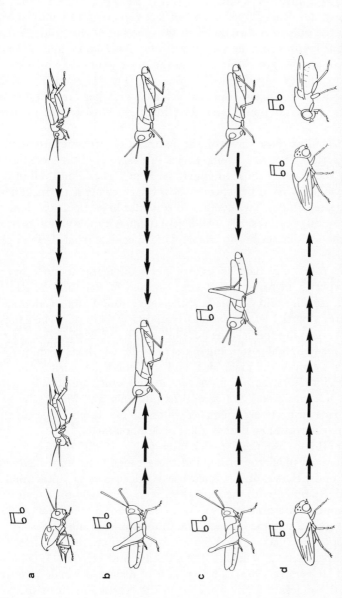

Figure 7.4. Acoustic strategies in insects. a, The male sings and the female goes to him. Found in most crickets, some grasshoppers and bush-crickets; b, The male sings, the female approaches and the male meets her; Grasshoppers and bush-crickets; c, The male sings, the female approaches and replies and the male is attracted to the female: Grasshoppers and bush-crickets; d, The male sings, the female replies and the male goes to the female: Gall midges and some bush-crickets. A large number of variations on these patterns of behaviour are found.

brief sounds produced by only a few of the teeth on the strigil being struck; 'lisps', pulses of sound in which there are a large number of toothstrikes, and 'clicks', consisting of between two and four short, closely spaced pulses. A general point to come out of this study is that the female responses to male song are always short 'ticks' and those of the different species are similar, while the male songs are much more varied and complex and characteristic of the species producing them. This indicates, once again, that the female, with her larger parental investment, exercises the greater discrimination with regard to mate choice.

Figure 7.4 illustrates some of the patterns of acoustic interaction between species. They vary from species in which all of the locomotion is done by the male, to those in which the female goes almost all of the way to the male, while other species show intermediate patterns. These different strategies may have evolved in response to particular ecological and behavioural pressures, but insufficient detail is known about the way of life of the various species to enable us to speculate what these might be.

In some cases the male songs are not sufficiently distinct to be adequate species identifiers. In three species of katydid, *Scudderia cuneata*, *S. furcator* and *S. fasciata*, the songs sound almost identical. While there is partial ecological and temporal isolation between these spatially overlapping species, the possibility of hybridization does exist. However, the temporal patterning of the male–female duets is unique, as the interval between the male 'lisp' and the female answering 'tick' is species-specific. For example, in the two species which are most likely to hybridize, *S. cuneata* and *S. furcata*, the intervals are 399 ± 4 ms and 1118 ± 25 ms respectively. Females which do not respond within the appropriate time window do not elicit male phonotaxis.

Spooner (1968b) lists the lisp–tick intervals for 13 different species of North American phaneropterine, and these range between 0.1 second and 1.4 seconds. He was able to make use of this feature to capture males in the wild. By replying to their songs with an artificially produced tick after the appropriate time interval he could attract males which called from inaccessible stations in thick undergrowth or on high trees.

Phaneropterine duetting has been investigated more recently in some European species by Robinson, Rheinlaender and Hartley (1986) and Heller and Helversen (1986). The patterns of male and female song are similar to those described by Spooner in the North American species. Most of the males produce a single wing closing movement resulting in a long pulse (= lisp) while females, also on the wing closing stroke, produce 'ticks' of less than one millisecond's duration (Figure 7.5). The response delay times between the onset of the male pulse and that of the female for six species is illustrated in Figure 7.6, and these vary from

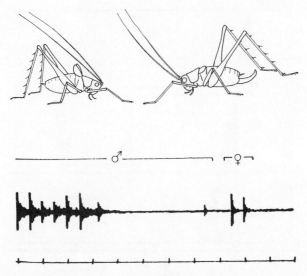

Figure 7.5. Oscillogram of the song of the Bush-cricket, *Leptophyes punctatissima*. Male song followed by the female response. Time marker, 5 ms. (From Hartley and Robinson 1976).

between 18 ms to almost half a second. If males are identifying females on the basis of these intervals then the males' response times should match the delays. Figure 7.7 illustrates the time periods over which males of two *Poecilimon* species will make phonotactic responses. The fit is clear except that the males also respond to longer intervals than might be expected.

In contrast to the songs of most other tettigoniids, those of phaneropterine males are very short. They consist of only a few ultrasonic pulses, while those of the females usually comprise a single pulse of less than one millisecond. This is a feature of these songs which facilitates accurate time measurement by providing an abrupt transient. However, in one species, *Ancistrura nigrovittata*, males produce pulse trains which are about 300 ms in duration. These are followed by single pulses and it is pertinent that the latter sounds are used as temporal cues by females when replying to male song (Figure 7.8).

One consequence of the short response times found in some of these species is that they can communicate only over a short distance. In *Leptophyes punctatissima*, for example, the female's acoustic response to male song occurs after a delay of only 25 ms. The male's response is confined to an 'acoustic window' of between 25 and 45 ms which allows a maximum of only 20 ms for the sound to travel to the female and for the reply to be received. Sound travels about one metre in six milliseconds which implies a communication range of only 3.3 metres for this species (Robinson, Rheinlaenader and Hartley 1986). This perhaps

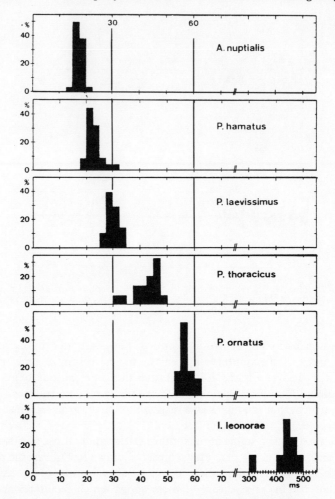

Figure 7.6. Distribution of the delay times of the female's response to male song in six barbitistine bush-cricket species. The delays were measured from the onset of the male's signal. (From Heller and Helversen 1986).

accounts for the observation that male response times are more spread out at the longer part of their range than are the female delay times, as this would increase the range over which the pair could communication. It should be borne in mind, however, that one would not expect communication over a long distance in insects which have ultrasonic songs. The fairly quiet ultrasonic calls of these bush-crickets are particularly subject to attenuation due to the shrubs and foliage in which they live.

It is tempting to conclude that the species-specific time delays and acoustic windows act as sexual isolating mechanisms, particularly as the

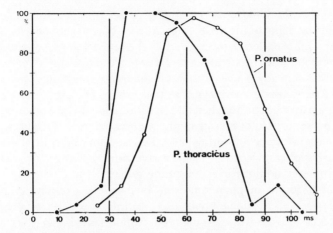

Figure 7.7. Response curves of the males of two *Poecilimon* species to synthesized female signals with different delays following male song. The female signals were played through a loudspeaker and the males' responses were expressed as percentage turns towards the speaker. (From Heller and Helversen 1986).

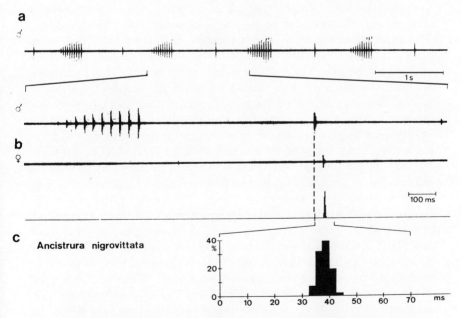

Figure 7.8. Duetting in *Ancistrura nigrovittata*. a, Male song showing the chirps, each followed by a single pulse; b, Female response to the single pulse of the male; c, Histogram of the time delay from onset of the male pulse to that of the female. (From Heller and Helversen 1986).

spectrum of the songs does not appear to be important for recognition. In addition, many of the male songs are too short to contain much useful temporal information. However, while the time windows may well serve this function they may have evolved in response to other selective pressures. One advantage of this type of communication system is that it has a particularly good signal to noise ratio. If a signal is 'expected' during a particular period then all sounds outside this are ignored, which could be important for insects living in noisy habitats. This is analogous to the situation in echolocating bats where auditory units in the brain are responsive only to echoes which arrive during a short time window following emission of the echolocating pulse.

What is found in phaneropterine bush-crickets is not unique and probably represents one extreme of a continuum of acoustic behaviour. Rupprecht (1976), for example, has demonstrated that the stonefly, *Sialis*, females require to reply to the drumming signals of males within 400 ms to elicit male phonotaxis. In the visual communication of fireflies (Lampyridae: Coleoptera) a similar phenomenon is observed and the correct temporal relationship of male–female flashes must be adhered to for phototaxis to result (Papi 1969, Lloyd 1977).

Another dimension is added to the phonotactic strategy of duetting insects by the behaviour of some bush-crickets, mainly *Ephippiger*, *Conocephalus* and *Copiphora* species, in which males produce both acoustic and substrate-borne signals (Dumortier 1963, Morris 1980). The function of the latter is not clear in *Ephippiger* but it has been observed that when calling males perceive the vibrations due to females arriving on the same plant, they produce tremulations. These probably act both to stimulate locomotion and to guide the females to them.

The phenomenon has probably been studied best in *Copiphora rhinoceros*, a large Central American species which lives in understory vegetation. *Copiphora* males perform a relatively complex song comprising a pulse train followed by a tone burst, the entire phrase having a duration of 0.5 to 1 second. This is repeated about 40 to 50 times a minute. Males also perform bouts of tremulation which consist of 2 to 5 pulses with a repetition rate of $1.3 \, s^{-1}$. The carrier frequency of the acoustic signal is 8.7 kHz with harmonics at 17.4 and 26.1 kHz. The characteristics of the vibrational signal do not appear to have been described. A simplified résumé of calling behaviour derived from field observations indicates that males produce alternate periods of stridulation and tremulation from a station on a plant. Males were observed to take up a position on the midrib of a leaf (*Philodendron* species), a position which would facilitate the transmission of vibrational signals. A female attracted onto the plant from which the male was signalling would respond with answering tremulations and approach the male.

This system, which makes use of two communication channels, has several potential advantages. The acoustic signal will be effective over a considerable range but will be subject to the disadvantages of such signals as described above. However, once the female answers by means of tremulations, the male can abandon the potentially dangerous loud calls in favour of a more intimate signal which will both guide the female and be relatively safe from interception by predators and competing males (Morris 1980).

CHORUSING AND AGGRESSION

Many arthropods aggregate for the purpose of mating. In insects such as cicadas and some of the Orthoptera, aggregated males perform calling songs in chorus. The choruses of cicadas are particularly conspicuous, with perhaps thousands of individuals congregating in a single tree to sing. Males come together in two possible ways. They may be attracted to the songs of conspecifics which could be termed active aggregation or, alternatively, they could collect at a particular location having a specific resource value. This, for example, could be a localized food plant or shelter from predation or the weather. These are 'resource based leks' (Alexander 1975) and are essentially passive aggregations of individuals. However, it would be false to consider these causes of aggregation to be necessarily mutually exclusive. It would be quite conceivable for phonotaxis to bring about aggregation at a particular resource.

While there are undoubtedly many examples of chorusing behaviour, there is little evidence demonstrating active aggregation of individuals. The requisite nearest-neighbour analysis which shows that the inter-individual distances of singing males are less than would be expected on the basis of random spacing, has been carried out successfully only on the Field Crickets, *Teleogryllus commodus* (Campbell and Clarke 1971), and *Gryllus* species (Cade 1981a) and the Bush-cricket, *Amblycorypha parvipennis* (Shaw, North and Meixner 1981).

The evidence for males being attracted to conspecific song is also scanty. The sound traps used by Dong and Beck (1981) for capture-release experiments on the mole cricket, *Scapteriscus acletus*, attracted males as well as females. About 20 per cent of the total consisted of males, with the proportion of the catch rising to 50 per cent on some occasions. Similar results were also obtained by Ulagaraj and Walker (1973) for *S. acletus* and for *S. vicinus*.

Even if we do not know the mechanisms whereby insect choruses come about, it is interesting to consider their adaptive significance. One of the functions of a chorus might be as an anti-predator or anti-parasite device. The former function has often been ascribed to bird flocks and fish shoals. A large number of individuals can distract and confuse potential predators. Animals on the periphery of such a group would be at risk, but those towards the centre would be at an advantage, and the overall shared risk would be less than that run by the individuals if they were on their own. This is the 'selfish herd' hypothesis of Hamilton (1971). In aggregations whose function is reproductive, as in insect choruses, individuals on the periphery might be at an advantage, as they would be in the best position to intercept an attracted female and mate with her, although this is not true of mammals and birds. Unfortunately the potential costs and benefits of such a system do not appear to have been evaluated for any insect.

The evidence for aggregation in the periodic cicadas, *Magicicada* spp., of North America, although less formal than for some Orthoptera, is nevertheless overwhelming. They come together in groups whose combined biomass exceeds that of any other terrestrial animal, with numbers of up to 300 individuals for each square metre (Dybas and Davis 1962). Enormous congregations of many thousands of singing males will crowd together in a single tree, and their songs attract not only females but other males, and the calls of the males stimulate others to sing. This positive feedback accounts in part for the concentration of animals (Alexander and Moore 1962). Karban (1982) has been able to demonstrate that reproductive success is positively correlated with local population density, showing that these aggregations of cicadas brought about by song are advantageous.

There are at least two possible reasons for this relationship between population density and reproductive success. First, the probability of finding a mate and mating successfully will be increased, particularly as these cicadas are poor fliers and attract many avian predators. Second, the levels of predation might decrease with increasing prey density. This has been shown to be true, for example, in chorusing tree-frogs (Ryan, Tuttle and Taft 1981). Estimating the degree of predation by counting the numbers of discarded cicada wings Karban was able to demonstrate that the same was happening in periodic cicadas and was due to predator satiation. A relevant factor in this system is the periodic nature of the cicada life-cycle where a long period, 13 or 17 years, depending upon species, passes before adult emergence. This precludes the build-up of the predators and parasites which might otherwise benefit from the bonanza when it occurred. This may well be the major selective force favouring the periodic life-cycle.

The above example is perhaps a rather special case, as the majority of cicadas, although they can have very long larval lives, are not periodic and do not have synchronized adult emergence. Other cicada species also form choruses, and in these, predator satiation is less likely to be one of the reasons for the aggregations. It is noteworthy that while the periodic cicada species are both poor fliers and rather brightly coloured, thus making them particularly liable to predation, other species fly well and are, even in their choruses, very cryptic.

The most obvious potential advantage of a chorus is that it acts as a lek; the pooled songs of several individuals would be more attractive to females than those of isolated males. They would be louder in absolute terms and might therefore attract females over a greater range. Most of the lekking behaviour that has been looked at in mammals and birds involves visual displays, but these are rare in insects, although notable examples are to be found in fireflies (Lampyridae)(Lloyd 1977), and Hawaiian *Drosophila* species (Spieth 1974). Once again, evidence for the greater attractiveness of communal song over the songs of individuals is limited, although Morris, Kerr and Fullard (1978) have shown that females of the tettigoniid, *Conocephalus nigropleurum*, choose the pooled songs of two males in preference to that of a single one. More *Gryllus intiger* females were attracted to aggregated speakers broadcasting conspecific song than to single speakers, although the average number attracted to single speakers within the aggregations was no greater (Cade 1981a).

ACOUSTIC INTERACTIONS BETWEEN MALES: MALE SPACING

Calling songs can be involved not only in attracting conspecific males and females but also in male spacing. While there are relatively few examples demonstrating the attractiveness of male song for other males, there are many more which show that songs are used to maintain the spacing between calling males. These two functions are not mutually exclusive and it is probable that in many cases calling songs both attract males from a distance and then dictate their local distribution. Spacing may thus occur within aggregations or choruses of males and also between males of solitary species. Thus, even in the widely dispersed bush-cricket, *Mygalopsis marki*, the spacing patterns show that both attraction and repulsion operate during calling. Females probably mature within the area occupied by males and are therefore not attracted to combined male song but rather to that of individuals within the aggregation. It is probable that in many cases calling is concerned with maintaining individual space or a territory. The acoustic interactions of males help to achieve this without the animals resorting to overt aggression. The calling songs are therefore distinct from the majority of aggressive or rivalry songs found in many species (see Chapter 8).

For convenience, rather than because any rigid demarcation exists, we can divide calling behaviour of males according to those species which form aggregations or choruses, and those which are solitary. Within the former group three different types of chorusing behaviour have been described: unsynchronized song, synchronized song and song alternation.

Unsynchronized chorusing

Unsynchronized chorusing occurs when males within a group call concurrently, with bouts of song interspersed with periods of silence. No apparent fixed relationship appears to exist between the individual songs and, although subtle interactions are possible, the pattern of individual song is obscured. Such choruses, while common, have been analysed in any detail for only one species, the bush-cricket, *Neoconocephalus affinis* (Figure 7.9a)(Greenfield 1983). The amount of song by individual males in the choruses was extremely variable, as were the durations of song bouts. The major conclusion from looking at pairwise interactions was that a positive feedback system was operating and, if one male was removed, the other would not sing. However, there were some persistent singers or leaders within aggregations who usually initiated singing. One possible interpretation of this behaviour is that it results from sexual selection. If one male started singing, it would be likely to gain the advantage in attracting a female and, so as to compensate, the others would have to join in. On the other hand, persistent singers, by singing earlier and for longer periods, would expend more energy and might be more liable to possible predation.

Other examples of unsynchronized chorusing are found in grass-hoppers, *Syrbula* species (Otte and Loftus-Hills 1979), bush-crickets, *Ambylocorypha* species (Alexander 1956) and cicadas (Alexander and Moore 1962). While some species of cicada sing together continuously for long periods of time, others, such as the *cassini* sibling species group of *Magicicada*, demonstrate spectacular unsynchronized chorusing. A few individuals will begin to sing and others join in in increasing numbers until the chorus reaches a crescendo which finally dies away to silence.

Synchronized chorusing

The tettigoniid, *Neoconocephalus nebrascensis*, provides a good example of synchronized singing. The males produce trills of 1 to 1.5 seconds' duration interspersed with silent periods of between 700 and 900 milli-seconds. Adjacent males produce trills in near synchrony so that the overall acoustic effect is of temporally patterned song. In pairwise in-teractions, one male, the leader, initiates the trill and the other, the follower, starts trilling 50 to 200 ms later. The follower also stops between 100 and 300 ms before the leader and therefore, as the two do not change the lead, followers sing less overall than do leaders (Figure 7.9b).

Experimental reduction of inter-individual distances between leaders and followers showed that the trill length of the follower became pro-gressively shorter, until at a critical point the follower stopped singing. In the field, locomotory movements sometimes bring males into close proximity, and here too follower males fall silent before eventually moving away. These songs therefore not only attract females but serve to space the males (Meixner and Shaw 1986).

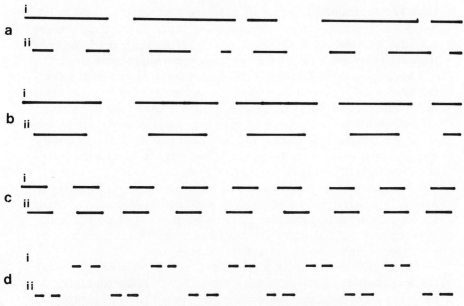

Figure 7.9. Patterns of interaction between pairs of calling males (i,ii). a, Unsynchronized song, e.g., *Neoconocephalus affinis*; b, Synchronized song with leader–follower relationship, e.g., *Neoconocephalus nebrascensis*; c, Synchronous song with changing lead, e.g., *Oecanthus fultoni*; d, Song alternation, e.g., *Pterophylla camellifolia*.

A rather different type of synchrony has been described in the Snowy Tree Cricket, *Oecanthus fultoni*. Fulton (1928) noted that chirping among individuals became asynchronous if they were deafened by destruction of the tympani, so demonstrating the means whereby the synchrony is achieved. Unlike what is found in *N. nebrascensis*, the temporal relationship between individuals is not fixed for long periods of time, but the lead alternates from one animal to another in a regular fashion (Figure 7.9c)(Walker 1969). These songs, while possibly concerned with spacing, are not therefore involved in dominance relationships.

Walker carried out experiments using tape loops of recorded song. These were slightly frequency shifted so that the responses of a singing insect could be distinguished from the taped song. Using this technique he could show that the synchrony between animals was maintained, not by the animals responding to the concurrent chirp on the tape for timing their own, but to the preceding chirp. The chirps are composed of between 2 to 11 pulses, although usually between 5 and 8, at 50 to 200 chirps per minute. Males are thus able to control both chirp repetition rate and chirp duration. The experiments showed that if the broadcast chirp commenced towards the end of the tree cricket's own chirp, or at the beginning of the subsequent interval, the insect delayed its own next chirp and sometimes increased its duration. On the other hand, if the broadcast chirp began towards the end of the interval, the cricket shortened its chirp and succeeding interval. The phase shifts resulting from these modifications improve the synchrony between individuals.

The responses of the Snowy Tree Cricket are in a sense anticipatory, unlike those of *N. nebrascensis* in which the timing of trills is based on the concurrent trill. The latter is only possible where the individual trills are sufficiently long for the recipient to be capable of processing the incoming acoustic information and responding to it. The anticipatory mechanism is essential in *O. fultoni* whose song is composed of very short chirps with a high chirp repetition rate.

Other examples of synchronous singing are in the Water Bug, *Sigara striata*, in which the phenomenon was identified by frequency shifting the song of one of the participants. It may be that the technical difficulty of demonstrating this type of singing has obscured its occurrence in some cases (Finke and Prager 1980).

One selective force which has given rise to synchrony is likely to be the preservation of the species-specific song rhythms, while at the same time providing louder and thus more attractive songs. That the maintenance of synchrony is an active process is emphasized by the experiments with a synchronizing bush-cricket, *Platycleis intermedia*, which attempts to retain the synchrony even when the insects are singing at different temperatures. Temperature is known to affect chirp rate dramatically in many species and the synchronizing mechanism would allow the calling animals to compensate for microhabitat variations in temperature (Samways 1976).

Another possible function of overlapping synchronized song pulses is suggested by the behaviour of the katydid (bush-cricket) *Ambylocorypha parvipennis* (Shaw, North and Meixner 1981), in which females reply to male calls with a brief 'tick'. A male who extends his chirp beyond that of his rival might be able to drown out the female's reply, and thus disrupt the obligatory fixed temporal relationship of the duet and so gain a possible advantage.

Song alternation

Alternation of chirps has been described in the calling songs of the grasshopper, *Chorthippus brunneus* (Loher 1957), and several species of bush-cricket, for example, *Pholidoptera griseoapterus* (Jones 1866), *Eph-ippiger* species (Busnel, Dumortier and Busnel 1956) and *Pterophylla camellifolia* (Shaw 1968). The latter provides a detailed example in which both the acoustic relationship between individuals and the effect of artificially propagated sounds were analysed. *P. camellifolia*, in common with other chorusing species, will also produce solo song. The chirp repetition rates of solo songs are higher than those of interactive ones. By making use of artificial chirps of controlled intensity, Shaw could show that hearing a chirp at above a sound intensity of around 55 dB had an inhibitory effect on the production of the next chirp. Its timing is therefore delayed, accounting for the longer inter-chirp intervals in alternating song. There is also evidence for a generalized excitatory effect of hearing the calling songs of conspecifics. Once one animal stops singing the effect on the other of having been involved in acoustic alternation is for individual chirps to become longer and for the chirp repetition rate to increase.

A computer model for alternating song in *P. camellifolia* has been developed by Soucek (1975), which is compatible with the behavioural data and with what is known of possible neuronal mechanisms. It is based on the interaction between the stimulus period (= input) and response or chirp period (= output) curves. A consequence of the interaction is that the resultant curve contains a relatively stable area where alternation of chirps with changing leader–follower relationships exists. Two abrupt discontinuities on the curve flip the behaviour towards solo singing on the one hand and partially delayed chirping on the other, both behaviours which are occasionally found.

P. camellifolia also alternates song during aggressive interactions, and the effect of alternation may be to space the participants. The transition from alternate calling to alternate aggressive song occurs when the inter-individual distance becomes too close. However, more effort has gone into describing this type of acoustic interaction than into attempting to elucidate its function.

It is not at all clear why different patterns of acoustic interactions have evolved, particularly as they can all be found in closely related species. Even within species more than one type sometimes occurs. Samways (1976) looked at five species of *Platycleis*, one of which, as described above, produced synchronized song, two alternated and the other two were asynchronous. They are probably all involved in spacing and inter-male competition, with sexual selection being the driving force. We will probably require more detailed comparative studies of behavioural ecology before an understanding is achieved of how the different acoustic strategies are tailored to the lifestyles of particular species.

Solitary singing

All of the above examples concern species which, to a greater or lesser extent, form aggregations. However, it is very difficult to make a distinction between aggregating and dispersive species. The extremes are obvious, but there is a grey middle area. In many cases two opposing forces are at work; calling song both attracts and repels others of the same sex, and the relative balance between these will be one factor which determines the degree of aggregation. With dispersed species it is difficult to know if their spread is limited by factors within the habitat or whether they occupy sites from which distantly calling conspecifics can still be heard.

Many bush-cricket species take up positions on trees and shrubs which may be many metres from the next calling male, and these are usually considered to be solitary. One such species which has been studied by Thiele and Bailey (1980) and Romer and Bailey (1986) is the West Australian *Mygalopsis marki*. In two study areas, the mean interindividual distances were found to be 6.2 m and 11.5 m. There are at least three different methods whereby these spacings could be achieved. The distances might represent the threshold for hearing, that is, males might space themselves out until just out of earshot. By the ingenious method of using the insect itself as a 'biological' microphone Romer and Bailey showed that this was not the case. They recorded from acoustic neurones responsive to conspecific song while moving the preparation around the study areas to measure the response to natural song under various conditions. The preparation could 'hear' conspecific song at double the usual spacing distance.

The song of *M. marki* contains two major frequency peaks at 10 and 20 kHz and, as the latter attenuates to a greater extent with distance, the relative intensities of these two frequency components could be used to measure range. This also was found not to be the mechanism used. The single factor influencing spacing was the overall intensity of the song and a sound pressure level of 65 dB determined the positions of the males. In the study area with 11.5 m spacing, there were small trees from which the bush-crickets called, and in this case there was less sound attenuation than in the other study area where the crickets sang from perches in low shrubs.

INFORMATION CONTENT OF CALLING SONGS

One question which has attracted much attention is the nature of the information contained in the calling songs which renders them attractive to conspecifics. Resolving the problem is made difficult because the calling songs may, as described above, have more than one function, and it is not always clear if males and females are responding to the same or to different song parameters. In addition, even closely related species may have quite disparate song patterns and it is often dangerous to generalize across species.

There are three different pieces of information which may be coded within calling songs. These are, range, position in space and species identity.

Location of sound sources

Range can be assessed on the basis of sound intensity and there is evidence that female orthopterans choose the louder, and therefore probably the closer, of two sound sources. In those species which produce calling songs with more than one spectral peak, the relative intensities of the different frequency components may also indicate range as the higher frequencies are selectively filtered by the environment. In an analogous manner, arthropods which use substrate vibrations for calling, such as fiddler crabs and many Hemiptera, could potentially utilize the relative loudness and timing of the different frequency components of a signal. This is because of the dispersive nature of substrate vibrations where propagation velocities are frequency dependent (see Chapter 1).

Receptors which respond to particle displacement such as the antennae of insects and the various hair receptors found in a wide range of arthropods are inherently directional. The direction of a sound can be located by integrating the input from different receptors. In theory it would be possible for an animal possessing only a single receptor to locate a sound by orienting itself to find the axis of maximum stimulus input.

With the more usual tympanal organs, direction of a calling song can only be discerned on the basis of the relative sound intensity and time differences at the two auditory receptors. These are designed so as to maximise the differences and, while most arthropods are too small for time differences per se to be measured, the response latencies of the receptor cells are intensity related. This effectively means that there will be relatively large time differences between the inputs from the two receptors reaching the auditory neuropile.

The ability of the auditory system in the grasshopper, *Chorthippus biguttulus*, to discriminate time and intensity differences has been investigated by von Helversen and von Helversen (1983). They looked at the orientation responses of calling males to the simulated replies of females. Two loudspeakers at 90° and 270° to the male, producing female calls whose relative intensity and timing could be controlled, were used to elicit male responses. The point at which the male's orientation responses became random coincided with an intensity difference of less than 1 dB or a time differential shorter than 0.5 ms (Figure 7.10).

There are a number of problems regarding the way in which phonotactically responding animals locate calling individuals, which have yet to be explored in detail. Most laboratory experiments use arenas or Y-mazes in which the problem is reduced to two dimensions. Similarly, neurophysiological experiments are normally done on restrained animals and the stimulus of the sound source is moved around the preparation in the horizontal plane. While this degree of control is essential in such experiments, many calling insects operate in an inhomogeneous, three-dimensional world possessing many more degrees of freedom.

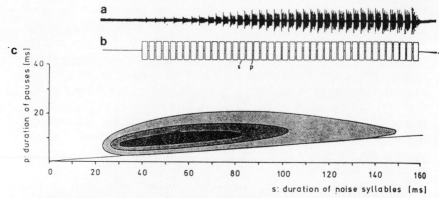

Figure 7.10. Parameters of the stimuli eliciting an acoustic response from female *Chorthippus biguttulus*. a, Oscillogram of male calling song; b, Artificial song used to stimulate females; c, The effects of changing syllable lengths (s) and inter-syllable-intervals (p). The contour lines correspond to 20%, 50% and 80% positive responses. The cross-hatched area shows the normal range of variation in actual male song. (From Helversen and Helversen 1983).

For an animal such as a cricket, with its ears on its forelegs, there are obvious difficulties in making binaural intensity measurements at the same time as walking. While in species such as *Teleogryllus oceanicus* or *Scapsipedus marginatus*, the animal pauses intermittently in order to make directional adjustments, *Gryllus* species are able to modify their orientation actively whilst walking and this is a much more difficult task (Murphey and Zaretsky 1972, Bailey and Thomson 1979, Rainlender, Shuvalov, Popov and Kalmring 1981).

There is some evidence to support the idea that the systems which are concerned with assessing the direction and recognition of calling song are functionally separated. Using a similar paradigm to that described above for measuring discrimination abilities the von Helversens scored female responses to simulated male sounds. The combination of two patterns, neither of which was effective in eliciting the response on its own, but when presented together mimicked male calling song, elicited female responses even if the sounds were fed separately into the two tympani. On the other hand, two patterns, each separately effective but when combined to produce an ineffective pattern, did not elicit a positive response if presented in this way. Thus the female was responding only to pattern and not the direction from which the sound came, even although the 'correct' pattern only came into existence within the central nervous system. By making a range of lesions in the connectives, it has been demonstrated that a single tympanic organ and its unilateral connections to the brain alone were required for a female to respond to male song. Directional responses, that is, turning towards the source, were only possible, however, if bilateral inputs to the brain were present (Ronacher, von Helversen and von Helverson 1986).

Pattern recognition

In discussing pattern recognition it is important to maintain a clear distinction between those parameters of calling songs which are both essential and sufficient for recognition, that is, the minimum requirements, and additional ones which may be involved in eliciting an optimum response. There has been a tendency in some of the earlier literature on phonotactic behaviour to confuse these two aspects of the signals and this, along with different techniques for assessing positive phonotaxis, has led to some conflicting results.

Calling songs can be characterized both in terms of their temporal and spectral features. The neurophysiological work on the processing of acoustic information described in Chapter 4 demonstrates very clearly that the spectral composition of calling songs can be a crucial prerequisite for recognition. This is particularly true in those species of cricket, bush-cricket and cicada whose calling songs contain sharply tuned carrier frequencies. Over and above the trivial point that the calling song frequencies must be clearly within the range of conspecific hearing, there are examples of both peripheral and central mechanisms of hearing which selectively tune the receptor system to the carrier frequency of calling song.

The importance of carrier frequency in determining the direction of the phonotactic response has also to be demonstrated. In *Gryllus* species and in *Teleogryllus oceanicus* low-frequency sounds around 4 to 5 kHz elicit positive phonotaxis. High frequencies, on the other hand, cause negative phonotaxis, even if the sounds are patterned in the same way as the calling songs (Popov and Shuvalov 1977, Moiseff, Pollack and Hoy 1978, Pollack, Huber and Weber 1984). In this particular circumstance the influence of carrier frequency overrides that of temporal parameters. The threshold values at which high- and low-frequency sounds elicit negative and positive phonotaxis respectively in *Gryllus bimaculatus* is illustrated in Figure 7.11. As is usual in sensory systems there is a trade-off between frequency and intensity which is particularly marked for the low-frequency channel. Sounds away from the best frequency still elicit positive phonotaxis if sound intensity is sufficiently high.

The songs of most insects are temporally patterned to some extent and insect acoustic systems are better adapted to analyse the amplitude modulations rather than the spectral composition of signals. It would be surprising therefore to find many examples where frequency alone was sufficient for recognition. However, one example where the spectrum may be of prime importance is in the grasshopper, *Omocestus viridulus*. Its stridulations consist of a series of syllables, made up from impulses, each the result of a single tooth strike. By using computer simulated calling songs in which both amplitude modulation and tooth impact rate could be manipulated, Skovmand and Pedersen (1978) demonstrated

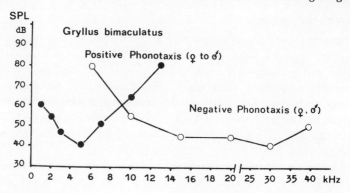

Figure 7.11. Threshold curves for the frequencies eliciting positive and negative phonotaxis in females of the cricket, *Gryllus bimaculatus*. Sound pressure level (SPL) is plotted against frequency. (From Popov and Shuvalov 1977).

that it was the latter which the females responded to. The impact rate changes in a stereotyped way during each syllable, and this is probably interpreted by the female as a frequency sweep. The major recognition parameter appears to be the frequency and not the amplitude modulation of the signal.

A good example of the variety of different temporal parameters of calling song involved in phonotaxis is provided by a comparative study of sympatric cricket species from southern Turkmenistan (Popov and Shuvalov 1977). They tested female phonotaxis in a Y-maze using both natural and synthesized calling songs as acoustic stimuli. Either a no-choice method was used, which demonstrated if a particular song pattern was or was not attractive, or females were presented with a choice of two patterns simultaneously to see which was preferred.

Gryllodinus kerkennensis males produce an almost continuous trill with poor separation of the individual pulses (Figure 7.12a) and females of this species respond equally well to an artificial tone burst of 5 kHz as to normal song. Synthesized song, in which the tone burst was broken into a pulse train with as little as 20 ms intervals between pulses, was discriminated against in a choice with continuous sound. This suggested it was the continuous nature of the calling song which determined its specificity. For *Gryllus bimaculatus* (Figure 7.12b) it appeared that pulse (i.e., syllable) repetition rate was the most important parameter. The number of pulses within chirps, normally between 3 and 5, and chirp repetition rates, were of minor importance. By contrast, in the closely related *G. campestris*, with very similar pulse repetition rate and carrier frequency, chirp duration, that is, the number of pulses within chirps, was much more important.

One undescribed species, 'Species 2' (Figure 7.12c), produces single,

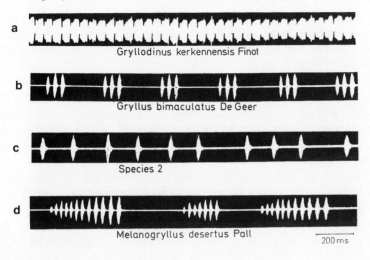

Figure 7.12. Oscillograms of songs of four sympatric cricket species from southern Turkmenistan. (From Popov and Shuvalov 1977).

short (26–32 ms) pulses which are repeated at rather large and irregular intervals. Pulse durations between 2.5 and 70 ms were equally effective in eliciting phonotaxis. As pulse length was increased beyond this figure, the song became progressively less attractive. Inter-pulse interval was more critical, with 80 ms intervals proving most effective. Finally, in *Melanogryllus desertus*, calling song is made up of amplitude modulated chirps which start quietly and become progressively louder (Figure 7.12d). Here the pattern of amplitude modulation as well as the pulse repetition rate were involved in eliciting phonotaxis, while chirp duration and interval were much less important.

These five examples demonstrate that quite different temporal parameters of song contribute to the essential recognition features in different species.

However, while the experiments identified certain salient features, they also showed that other parameters, while not essential for recognition, did affect the performance of phonotactic behaviour. Results from the choice experiments led Popov and Shovalov to conclude that the closer a song model was to the normal calling song of the species, the more attractive it became. In addition, the intensity of the female's phonotaxis as measured, for example, by the latency of her response to song, was also influenced by these non-essential parameters. They therefore divided calling song features into 'essential recognition' parameters and 'motivational' parameters. It could be that these two aspects of the signal are assessed by different mechanisms within the nervous system, possibly in a sequential fashion. Once song recognition has occurred and a female starts to approach a sound source, a song model which would be quite inadequate to initiate phonotaxis is often sufficient to induce her to continue. This is a commonly observed phenomenon in insects.

Recently research has been carried out on phonotactic behaviour, mainly of *Gryllus* species, using the Kramer spherical treadmill. This allows a precise measure of phonotaxis over a long period of time, during which a complex stimulus paradigm can be used. Subtleties in behaviour can be revealed by this method which are not apparent in, for example, experiments with Y-mazes.

Thorson, Weber and Huber (1982) demonstrated that the essential feature of calling song required to elicit phonotaxis or tracking in *G. campestris* was a 5 kHz carrier frequency amplitude modulated at 30 Hz, the syllable repetition rate of normal song. Although calling song consists of 4-syllable chirps, they found that trills were equally effective. The duty cycle, that is, the ratio of sound to silence, was also unimportant. While in real song this is about 50 per cent, artificial songs with duty cycles between 10 and 90 per cent also elicited tracking. These results are consistent with those of Popov and Shuvalov described above in so far as they implicate syllable repetition rate as being the essential recognition parameter in this species.

The Kramer treadmill was also used by Doherty (1985a) to examine phonotaxis in *G. bimaculatus*. He made use of an experimental paradigm where the female had to choose between different synthetic song patterns presented sequentially to the two sides of the animal. Left–right tracking was automatically scored while the stimulus parameters were systematically varied. He found that females would track songs with syllable repetition rates of between 20 and $30 \, \mathrm{s}^{-1}$ thus confirming the results of earlier workers. However, if the variability of the female phonotaxis was looked at rather than her all-or-nothing response, then the preferred range was much narrower at between 22 and $25 \, \mathrm{s}^{-1}$. The use of this criterion suggested a much more precise optimum value than previously supposed.

To assess further the relative importance of different song parameters Doherty (1985b) presented females sequentially with two songs, each having a single parameter varied, while the remainder were kept constant. In this way he could, for example, compare the effects of different syllable periods with different chirp periods while keeping the durations of both syllables and chirps the same (Figure 7.13). What such experiments demonstrated was a 'trade-off' phenomenon whereby the animal evaluated most of the song parameters simultaneously. If one of these is at or near the optimum value for the species, then another is permitted to deviate from its optimum without greatly affecting phonotaxis. The weightings of the parameters differ, as would be

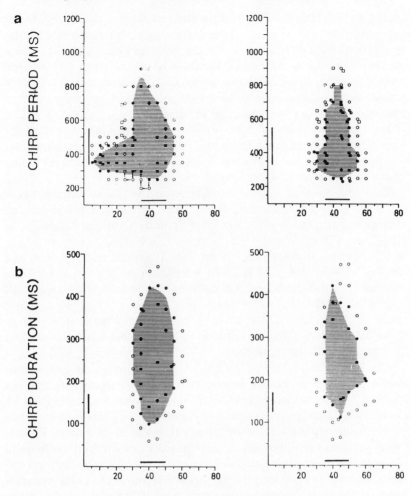

Figure 7.13. An example of the 'trade-off' phenomenon in the phonotactic behaviour of the cricket. a, Syllable period and chirp period are varied, while all other song parameters are kept constant. b, Chirp duration and syllable period are varied. Responses of two different females are illustrated. Open symbols = No tracking; closed symbols = Positive tracking; the bars indicate the range of variation found in normal song. (From Doherty 1985b).

expected, with essential recognition features, such as syllable repetition rate, having a high weighting. In other words, absolute values for attractiveness could not be ascribed to individual parameters, as this depended upon how close other song features were to the optimum.

Using a quite different experimental technique, Stout, DeHann and

McGhee (1983) analysed the relative attractiveness of the different song parameters in the cricket, *Acheta domesticus*. They plotted 'polar orientation' diagrams based on the direction and distance moved by a female in response to a broadcast synthesized calling song. This allowed them to quantify the female's responses. Each of the parameters, syllable period, syllable duration, number of syllables per chirp, chirp rate and frequency spectrum was evaluated in this way.

The clear conclusion of this work was that females assessed all of the parameters, although not with equal weighting. The strongest phonotactic response was to a song in which all parameters were close to the modal values of the natural calling song.

All of the research summarized above on the role of temporal parameters of calling song has been carried out on crickets. The one detailed analysis outside the gryllids has been done by Otto and Dagmar von Helversen (1983) using the grasshopper, *Chorthippus biguttulus*. Males of this species produce long chirps which contain many pulses, and females answer with a brief chirp which can be used as a response measure. Natural songs have a wide frequency spread, and females will respond to synthesized songs comprising pulses of 'white' noise whose spectrum ranges from 2 to 40 kHz. Various temporal parameters such as stimulus duration, syllable length and inter-syllable-interval, can then be experimentally manipulated.

Von Helversen and von Helversen (1983) showed that chirps with minimum durations of one second were essential to elicit female responses. The relationship between syllable duration and inter-syllable-interval within chirps is illustrated in Figure 7.10c. There is no evidence in this instance that a trade-off phenomenon is in operation. Rather, there is a fixed ratio between the two measures which renders the song recognizable even when these parameters are considerably beyond the limits found in normal song. Nevertheless, the durations of the intervals between the syllables are much more critical than the lengths of the syllables themselves. Interval durations are ultimately limiting and, if these exceed 35 ms, the song fails to elicit a response (Figure 7.14b). Rather unexpectedly, if the interval is interrupted by a brief (5 ms) pulse of sound, the efficacy of the song is restored (Figure 7.14c).

The individual syllables are amplitude-modulated and start with a few louder tooth strikes. Artificial sound pulses with this pattern of modulation are more effective than are those with a mirror image pattern of sound intensity.

Butlin and Hewitt (1988) have used the criterion of male mating success to assess the relative importance of song parameters in *C. brunneus*. Although this meant that they could only make use of the range in variation found in natural populations, the technique has a behavioural relevance lacking from some of the more experimental methods that

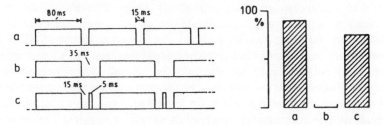

Figure 7.14. Responses of female, *Chorthippus biguttulus* to artificial song syllable. a, An optimum signal is composed of 80 ms pulses with 15 ms intervals; b, Increasing the interval to 35 ms abolishes the response. If, however, the longer interval is interrupted by a brief (5 ms) sound pulse as in c, the response is restored. (From Helversen and Helversen 1983).

have been employed. They showed that males with syllable lengths near the species mean were preferred by females. A Fourier transform of males' song revealed a previously undetected trisyllable structure that also conferred an advantage to males.

Experiments similar to those carried out with crickets to determine the relative importance of the different song parameters have not been done to the same extent with grasshoppers. Nevertheless, the results that are available indicate that the recognition system is, as in crickets, a complex one. It would be unsurprising if it transpired that the most effective artificial songs were those which resembled the natural ones in most of their features.

Conclusion

The distinction between essential and optimal acoustic parameters of calling songs is clear from the experiments that are described above. The relevance of this distinction lies mainly in the use that is made of the information. If one is concerned with identifying the neuronal circuits and elements involved in song recognition, then it is more practical to concentrate upon the essential features. The success of this approach is obvious from the results presented in Chapter 4, such as the identification of brain cells with response curves for song parameters which parallel those obtained from phonotactic tests.

With behavioural problems, such as the possible role of song in sexual selection, it may be important to take all of the parameters into consideration. In addition, it should be remembered that acoustic stimuli are not the only ones which may be involved. Vibrational stimuli will elicit positive phonotaxis in *Gryllus bimaculatus* (Weidmann and Keuper 1987), and we know that vibrational and acoustic stimuli are co-processed in the central nervous system of orthopterans.

Almost all of the work on phonotactic behaviour has been done on various orthopteran species, in particular crickets. It may not always be valid to extrapolate to other arthropods where different stimuli and sensory modalities could operate. Males of the Caribbean Fruit Fly, *Anastrepha suspensa*, produce a calling song by wing vibration, while at the same time releasing an attractant pheromone. Each of these stimuli can elicit phonotaxis in females, but they usually act synergistically (Webb, Burk and Sivinski 1983). The periodic cicadas intersperse bouts of singing with synchronized flight behaviour, and both acoustic and visual stimuli are probably involved in attracting females. This is true also of the crepitating flights of some grasshoppers (Riede 1987). There may well be trade-offs, not only between different calling song parameters, but also between different sensory modalities.

8. Song in Courtship and Aggression

COURTSHIP

Once contact between the sexes has been established, males may produce courtship songs. In some species of arthropod, females also sing and duets are performed. There is much less evidence on the possible functions of courtship song than there is for calling. In general terms, all courtship behaviour helps to ensure that mating occurs between conspecifics which are in the appropriate physiological state while, at the same time, providing the basis for possible mate selection. Courtship song contributes to these processes. Calling also can be considered to be part of courtship, and in the following discussion on function I refer, where relevant, to calling song as well as courtship song.

There are several different ways in which calling and courtship songs are involved in mating behaviour. Some tettigoniids, acridids and cicadas, for example, perform only calling songs, and these act so as to bring the sexes together. The subsequent courtship can be very brief or it may involve complex pheromonal, tactile and visual stimuli. In the majority of gryllids and in some Hemiptera and Coleoptera, calling songs are followed by courtship songs. These two songs are usually quite distinct from one another, but this is not always the case (Alexander 1962, Jansson 1973, Ryker 1976). In general, courtship songs are more variable than calling songs. The latter are instrumental in achieving species recognition and this constraint is therefore removed from courtship song.

In *Gryllus campestris*, for example, calling song consists of chirps containing syllables whose repetition rate is critical for eliciting optimum phonotaxis. Courtship song, by contrast, is made up of irregular pulse trains of rather soft sounds with irrregularly spaced, louder pulses (Figure 2.17). In crickets and many other insects the courtship songs are quieter as they are involved in close-range communication. The courtship signals therefore are less likely to be intercepted by rival males or by possible predators. In addition, in *Gryllus* species, although not in all crickets (e.g., *Teleogryllus*), the carrier frequency of courtship song is higher, 16 kHz as opposed to 4 kHz for calling song. The high-frequency sounds are more liable to attenuation, which also places limits on the range of courtship song (Figure 8.1).

While most attention has been paid to the Orthoptera and Hemiptera (Cicadas) because of their loud calling songs which may or may not be

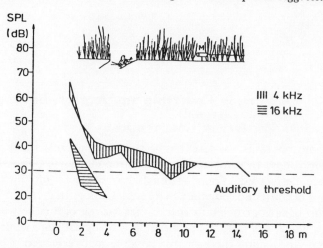

Figure 8.1. The audible range (in metres) of the major frequency components of calling (4kHz) and courtship (16kHz) songs in the natural habitat of *Gryllus campestris*. The cricket is singing at the mouth of his burrow and the microphone (M) is placed at ground level. The auditory threshold is taken to be 30 dB. Cross-hatched areas indicate the range of variation in the measurements. (From Popov, Shuvalov, Svetlogorskaya and Markovich 1974).

accompanied by courtship song, many insects have only courtship songs. They use ecological cues or alternative behavioural mechanisms, such as chemical or visual signals, to come together to mate. Of these, chemoreception involving the use of pheromones is probably the most widespread method. Ecological cues in the form of specific chemical components in food plants are used by many herbivorous insects, such as the Auchenorrhyncha (plant bugs and hoppers), while some Diptera use topological markers to form mating swarms (Downes 1969).

The majority of small insects that communicate acoustically can do so only at short range because of particular biophysical limitations. These are discussed in Chapters 1 and 2. The Auchenorrhyncha have avoided this problem to some extent by making use of substrate-borne signals rather than acoustic ones. The superior sound transmission properties of water allows some of the small aquatic Hemiptera to communicate at a distance which has been shown to be as great as half a metre for *Corixa* species (Jansson 1973).

Temporal synchrony in courtship

The role of song in limiting courtship to a physiologically appropriate time period is easy to appreciate. Males do not start to sing until they are capable of reproduction. This may be a matter of some hours after eclosion, as in some *Drosophila* species, to a few days in larger insects like the Orthoptera. Crickets, for example, do not start to produce calling songs until they have manufactured a mature spermatophore several days following the final moult. Proprioceptive feedback from the male reproductive system is responsible for the release of inhibition to sing. The neuronal circuits controlling song are, however, complete and potentially functional as early as the final nymphal instar (Bentley and Hoy 1970).

Female arthropods do not usually mate until their ovaries are fully developed. In grasshoppers and crickets this does not occur until four or five days following the moult to the adult stage. In several insect groups, the switch from an unreceptive state to copulatory readiness has been shown to be mediated by the corpora allata. Excision of the corpora allata from immature *Gomphocerhippus rufus* renders females permanently sexually unreceptive (Loher and Huber 1966). In *Drosophila* also, female receptivity is associated with juvenile hormone levels and this hormone is produced by the corpora allata (Manning 1967b).

Sexually unreceptive females of the House Cricket, *Acheta domesticus*, become receptive following implantation of corpora allata from receptive females (Stout, Gerard and Hassa 1976). Furthermore, this procedure has the effect of making females respond phonotactically to conspecific calling songs to which they were previously unresponsive. Immature and therefore sexually unreceptive *A. domesticus* females are actually repelled by sounds over a wide range of frequencies between 2.5 and 45 kHz. Positive phonotaxis to conspecific calling song, which has a carrier frequency of 4.5 kHz, is not seen before five days following the imaginal moult when they mature (Shuvalov and Popov 1971).

A well-known example of acoustically mediated mating synchrony is provided by the mosquito, *Aedes aegypti*. Males are attracted to the flight tone of sexually mature females. At eclosion, the flight tone is below the threshold of male hearing. The wing beat frequency and thus the flight tone becomes progressively higher after eclosion. The time of female sexual maturity coincides with the period when the flight tone has increased in frequency to fall within the range of male hearing (Roth 1948). In another species *Anopheles quadrimaculatus*, the acoustic transducer, the antenna, is plumose. The individual fibrillae on the antennae are only erected when the male genitalia rotate. This process is necessary for successful copulation and occurs between 15 and 24 hours after eclosion. Immature males are therefore effectively deaf to the female flight tone until ready to mate.

There are thus anatomical and physiological mechanisms which affect the timing of reproductive behaviour. The involvement of the corpora allata in this process in insects facilitates the integration of the physiological and behavioural components of courtship.

Sexual stimulation and reproductive strategy

While the courtships of some arthropods appear to be both very simple and short and have sometimes been described as rape by the male, the courtships of others are prolonged and complex. The former are often associated with reproductive strategies where males either hold significant resources or where they have gained the opportunity to mate through male–male competition. Where males do not hold resources, longer courtships may be required for females to be able to assess the suitability of a potential mate.

It has often been assumed that females require some specific amount of courtship stimulation before becoming sexually receptive. Earlier workers tended to use terms like female 'coyness' which male courtship was supposed to overcome. Latterly it became more fashionable to invoke sexual thresholds which required to be lowered. However, it is difficult to envisage precisely what is meant by either of these concepts. A more useful way of looking at the problem is to suppose that females are estimating male fitness either through the persistence, quantity or quality of his courtship. In this context it would be most interesting to know which components of acoustic behaviour are correlated with fitness and how females assess these.

Extended courtship may be considered best in relation to sexual selection. A female may possess an absolute criterion which a male must reach in his courtship, alternatively, a female might compare the courtships of competing males. There has been renewed interest in this issue in recent years, but the study of acoustic behaviour in arthropods has contributed rather little to date, although the potentiality for doing so is there (Blum and Blum 1979, Bradbury and Andersson 1987).

In evolutionary terms there is a conflict of interest between males and females which might account for the alternative reproductive strategies that are found. Singing is energetically expensive and potentially dangerous and it is therefore adaptive for a male to sing as little as possible during courtship or even not to sing at all if possible. In some crickets and cicadas there are non-singing satellite males which intercept and mate with females attracted to calling males (Cade 1975, Dunning, Byers and Zanger 1979). *Drosophila melanogaster* males normally perform a very brief courtship bout containing a single song burst on encountering a female. This is sometimes successful but is more often followed by an extended period of courtship before the female either accepts or rejects him. There is no clear indication what the difference is between these two types of courtship.

In gomphocerine grasshoppers the courtship sequence involves calling by the male which is answered by the female and a duet ensues. During this the male approaches the female and, sometimes, the female also moves towards the male. On contact, the male performs his complicated courtship ritual. This includes head, antennal and body movements as well as bouts of courtship song which are answered by the female until, finally, copulation is achieved. However, in the wild, males move around and do not call from a stationary position as do most crickets and bush-crickets. They frequently meet females and start courting without going through the calling phase, and the females may not reply to their courtship songs. This latter type of courtship is very much longer than the ones in which females participate acoustically.

Loher and Chandrashekaran (1972) observing *Syrbula fuscovittata* in the field and in the laboratory, only noted female song in the latter circumstance. Riede (1983) has shown that *Gomphocerus rufus* females go through several distinct phases of behaviour following the final moult. Newly emerged females up to about four days of age are totally unreceptive sexually. During days 5 and 6 they are in a state of 'passive acceptance' when they do not sing and only mate following extended male courtships. Females not mated during this period pass into an active phase when they reply to male song and copulate after the minimum of courtship. There then follows an unreceptive period when the female lays her eggs and a similar cycle ensues. There will be a trade-off between the need to select the most fit male and the possibility that by being too choosy the female might fail to find a mate at all. It will therefore be advantageous for a female to be highly selective initially and become progressively less so the longer she remains unmated. This consideration probably accounts for the changing female behaviour that is observed over time.

It is also worthwhile for the male to persist in his courtship even if the female is initially unreceptive. Riede has shown that one of the effects of male courtship on an unreceptive female is to inhibit her locomotion so that she does not move away from him. This inhibition of the female's tendency to escape is probably widespread. It has been commented on in, for example, *Drosophila* species, the bug, *Ligyrocoris diffusus* and the beetle, *Hylobius abietis* (von Schilcher 1976a, Thorpe and Harrington 1981, Selander and Jansson 1977).

In gomphocerine grasshoppers of the genus *Chorthippus*, Butlin and Hewitt (1986) have shown that female response stridulation and female phonotaxis are most commonly found in virgin animals which have reached the stage of having to lay unfertilized eggs if they remain unmated. Population density is an additional factor in deciding whether long or short courtships are performed. At high densities, males are likely to find females without having to call, and long courtships will occur. Only at low densities are females unlikely to have come across a male during the passive acceptance phase.

The importance of courtship song in inducing female receptivity may vary within species, depending upon ecological factors as is found in grasshoppers. There is also considerable variation in the importance of courtship song between species, even closely related ones. Thus in lacewings, *Chrysopa* species males and females exchange vibrational signals, but this does not appear to be an essential precursor of mating. In the related *Chrysoperla* species, on the other hand, these courtship signals are obligatory (Henry 1983).

Male mating success appears to be little affected by the presence of courtship song in some species of *Drosophila* and *Zaprionus*, chalcid wasps and the bug *L. diffusus* (Lee 1986, van des Assem and Putters 1980, Thorpe and Harrington 1981). However, in some of these examples the relevance of courtship song may be revealed only by observation under natural conditions. Moreover, it is unsafe to conclude that because song occurs during courtship its primary function is to stimulate females. It has been suggested, for example, that in *Zaprionus tuberculatus* the song is used to identify gender. It possibly acts to inhibit homosexual courtships which otherwise tend to occur in crowded conditions (Lee 1986).

At the other extreme, song in some insects seems to be an obligatory, or at least major, component of courtship. In the Pine Weevil, *Hylobius abietis*, and water bugs of the genus *Sigara*, mating is never observed in the absence of male courtship song (Selander and Jansson 1977, Jansson 1975). While visual rather than acoustic courtship signals are of major importance for certain *Drosophila*, in the most studied species, *D. melanogaster*, song is extremely important (Ewing 1983). It has proved possible to quantify the role of courtship song by progressively amputating portions of the wings and thus reducing the intensity and range of the song. The reduction in male mating success indicates that normally around 80 per cent of sexual stimulation is provided by wing vibration (figure 8.2). Totally silent males do eventually succeed in mating and antennectomized, and therefore deafened females, will also accept a male, demonstrating that song is not an essential courtship component (Ewing 1964, Manning 1967a).

Males have been observed to sing both during and after copulation. The former has been described in *Zaprionus* and *Drosophila* species, some beetles and bugs (Bennet-Clark, Leroy and Tsacas 1980, Selander and Jansson 1977, Ryker 1976, Heady, Nault, Shambaugh and Fairchild 1986). Many species of insect also produce tactile and contact-vibrational signals during copulation. These are outside the remit of this book, but it seems probable that the main function of all of such stimulation is to keep the female quiet during sperm transfer. No experiments appear to have been carried out to test this possibility.

The function of post-copulatory song as found in *Zaprionus* and passalid beetles is also unclear (Schuster 1983). It is possibly a form of mate-guarding behaviour which stops females from being remated, thus preventing sperm replacement. *Zaprionus* are often found in dense aggregations and, although females become unreceptive following mating, under these conditions a second copulation might be possible for

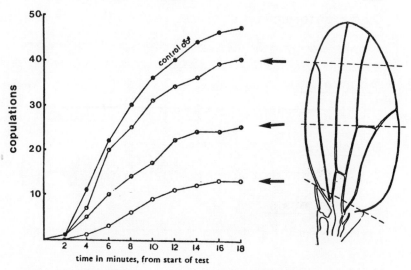

Figure 8.2. The relationship between mating speed and wing area in *Drosophila melanogaster*. The positions at which the wings were amputated is illustrated. The graph plots the number of copulations out of 50 pairs of flies over time for fully winged control males and those with wings amputated to different extents. (From Ewing 1964).

a brief period following dismounting. Some evidence for such a function exists in the solitary bee, *Centris pallida*, where females which have not been exposed to post-copulatory song remate more often (Alcock and Buchmann 1985).

SEXUAL ISOLATION AND SEXUAL SELECTION

Experiments designed to assess the ability of females to discriminate song parameters show that a very sophisticated level of information processing has evolved in crickets and grasshoppers (see Chapter 4). This is particularly marked for temporal features of song. There is every reason to suppose that when the appropriate experiments are carried out with other groups of arthropods complex acoustic discrimination will be demonstrated and the potential for females to choose mates on the basis of their songs confirmed. They could make use of the information contained in calling and courtship songs to recognize conspecific males. Evidence for the role of calling song in promoting species recognition is overwhelming, and some of it has been presented in Chapters 4 and 5. Females might also be able to use the songs for mate choice; that is to say, song could have a part to pay in sexual selection.

In spite of the fact that Charles Darwin (1871), in propounding his ideas on sexual selection, listed insect song among the characters which might be subject to such selective forces, this aspect of song has been almost ignored until relatively recently. Much more attention has been

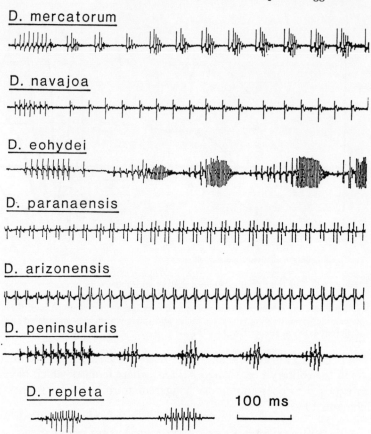

Figure 8.3. The species within the *Drosophila repleta* group have very different song patterns. Howeve4, almost nothing is known of the extent to which these differences contribute to sexual isolation.

paid to the role of song in sexual isolation than in sexual selection. During the course of evolution both of these factors are likely to have influenced the nature of song, and it is difficult if not impossible to separate their relative importance. Nevertheless, it is convenient to consider these two topics separately.

Sexual isolation

The fact that many arthropods have calling songs which preferentially attract conspecifics of the opposite sex or elicit acoustic responses from them, demonstrates that the songs are concerned in the initial stages of premating isolation. At close range there are likely to be other reinforcing isolating mechanisms based on visual displays, pheromones, tactile or even additional acoustic signals. Sound traps designed for one particular species often attract some individuals of other species. Equivalent mistakes must occur naturally and this will necessitate additional methods of species identification. In addition, in mobile animals such as many grasshoppers, the sexes often meet, court and mate without the preliminary of calling song.

There is a distinction to be made between species recognition and sexual isolation. Most, if not all of the calling songs produced by arthropods, have some part to play in species identification, and are therefore potentially capable of effecting sexual isolation. However, the songs will only effect premating isolation if sympatric species are present and there is the possibility of actual crossmating.

While there are many examples of groups of related species in which the individual species have distinct songs, often insufficient is known about their distribution and ecology to assess the effectiveness of the songs as isolating mechanisms (Figure 8.3). Even if species are sympatric, they may be separated by life history or ecological factors. The North American cricket species *Gryllus veletis* and *G. pennsylvanicus* are effectively isolated because the former overwinters as a late stage nymph, becoming adult in the spring or early summer. *G. pennsylvanicus*, by contrast, has a diapausing egg and the adults do not mature until late summer. Although some overlap of the adults occurs at the southern part of their range and both species have similar calling songs, hybrids are never found. Even if mating is induced in the laboratory, no offspring are produced (Alexander and Bigelow 1960). Temporal separation along with post-mating isolating mechanisms have presumably been sufficient not to require the evolution of different calling songs. The related, partially sympatric and synchronic species, *G. ruber*, has a quite distinct calling song. This suggests that in this species song has diverged as an isolating mechanism.

Some phytophagous insects, such as the smaller Hemiptera, with their substrate-borne calling songs, have strict associations with particular plant species or genera which give rise to an almost complete ecological isolation, while other species are generalists. One might predict that the latter would have more diverse calling songs, but this aspect of their acoustic behaviour does not appear to have been studied.

A few examples of closely related, potentially hybridizing species have been studied whose calling songs are likely to be important in maintaining sexual isolation. Members of the complex *Allonemobius fasciatus* group of crickets from the eastern United States, can be separated one from another only on the basis of electrophoresis or their songs. They have overlapping geographical distributions and the songs are undoubtedly one of the isolating factors. Two of the species, *A. fasciatus* and *A. socius*, do form inter-specific hybrids where their ranges meet, and it is perhaps significant that their calling song patterns are the most similar (Howard and Furth 1986).

Sexual isolation between the two grasshoppers, *Chorthippus brunneus* and *C. biguttulus*, is maintained by pre-mating mechanisms which include differences in song. Hybrids between the two species are obtained easily in the laboratory, but are not observed in the wild, although the species are sympatric over a large part of their range (Perdeck 1958). The ground crickets, *Nemobius allardi* and *N. tinnulus*, are also interfertile, and do not interbreed in the wild and they too have different songs (Fulton 1933).

The possibility of interspecific courtship and mating is potentially greatest where several different species are in close proximity. This situation is not often encountered, due to ecological factors such as niche diversification. It does occur in drosophilids where several species can congregate on a localized food source such as a piece of rotting fruit. It has also been observed in aquatic bugs, where several *Cenocorixa* and *Sigara* species may share a relatively small body of water. The species-specific pattern of male–female calling heard in some of these insects may be a mechanism which has evolved to help maintain isolation under these conditions. It is interesting from the sociobiological point of view that in the duets of water bugs, males are much less discriminating than are the females. Males will respond to the songs of heterospecific females and even sometimes to those of other genera. Females, however, respond only to the calls of males of their own species (Jansson 1973).

In those insects which perform both calling and courtship songs it is possible that the latter may reinforce the isolating role of calling song. Courtship songs do differ between species but much less work has been carried out with courtship as compared to calling song, and experimental evidence on this point is lacking.

The best documented examples of courtship songs whose function is at least partially concerned with sexual isolation are in fruit flies of the genus *Drosophila*. The songs of over one hundred species have been described to date, and an enormous range of patterns has been found (Cowling and Burnet 1981, Ewing 1970, 1979c, Ewing and Bennet-Clark 1968, Ewing and Miyan 1986, Hoikkala, Lakaovaara and Romppainen 1982).

Of particular interest are those examples where the songs of closely related, sympatric species have been compared. One which has been extensively studied by population geneticists is the sibling species pair *D. pseudobscura* and *D. persimilis*. Each of these species performs two different courtship songs, a low repetition rate and a high repetition rate song. There are differences between the species in the relative proportions of the two types of song, their repetition rates and carrier frequency (Figure 8.4) (Ewing 1969, Waldron 1964). While these two species, which are sympatric over a large area of North America, will hybridize in the laboratory and produce fertile offspring, interspecific hybrids have

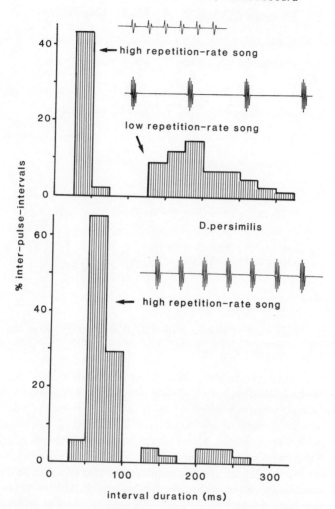

Figure 8.4. Songs of two sibling *Drosophila* species, *D. pseudoobscura* and *D. persimilis*. The former species performs two distinct songs, a high repetition rate, monocylic pulse song and a slower, polycyclic one. *D. persimilis* has one song only, a high repetition rate, polycyclic song. (From Ewing 1969).

not been found in the wild. The extent to which ecological factors and behavioural isolating mechanisms other than song are involved in sexual isolation has not been ascertained. That two sympatric species, which are almost indistinguishable morphologically, have such distinct songs is strong indirect evidence that song acts as an isolating mechanism.

The species which make up the *melanogaster* subgroup of *Drosophila* species also differ markedly in their courtship song patterns (Figure 8.5).

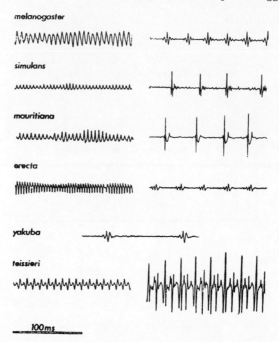

Figure 8.5. The songs of sibling species of the *D. melanogaster* group. All of the species, except for *D. yakuba* have two songs, sine songs and pulse song. (From Cowling and Burnet 1981).

This group originated in the African continent and some of the neighbouring islands, although two sibling species *D. melanogaster* and *D. simulans* are now cosmopolitan. In their original habitats it would have been common for more than one of these species to be found together. In East Africa, for example, I have observed three of the species, the two cosmopolitan ones and *D. yakuba* along with *Zaprionus* species on the same food source. Under these circumstances one can see much interspecific courtship. Even although the courtship behaviour of these species has been subject to detailed analysis, the extent to which the songs contribute to the isolation between them is still not clear (Cobb, Connolly and Burnet, 1985).

The Planthopper, *Nilaparvata lugens*, a major pest of rice in Asia and Australia, has been shown to comprise two distinct biological species. One feeds on the grass, *Leersia hexandra*, and the other on cultivated rice. While these races will cross in the laboratory, no naturally occurring hybrids have been observed. The pulse repetition rates of the substrate-transmitted songs have been shown to be one of the factors responsible for maintaining sexual isolation, and these differences are particularly clear-cut between sympatric Australian populations of the bug (Figure 8.6) (Claridge, den Hollander and Morgan 1984).

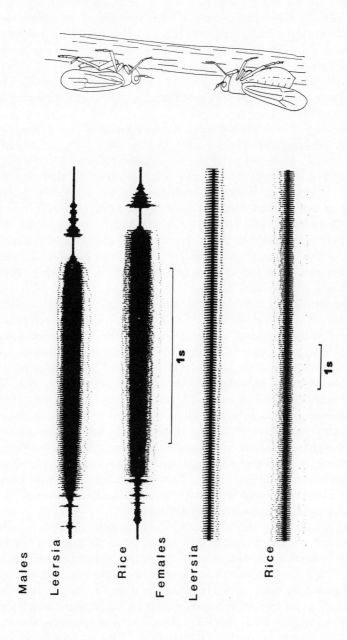

Figure 8.6. Vibratory songs of members of two sympatric races of the planthopper, *Nilaparvata lugens*, from northern Australia. Both sexes of the *Leersia* population have lower pulse repetition rate songs than those from rice. (From Claridge, den Hollander and Morgan 1988).

As described above, *D. melanogaster* males silenced by amputating their wings are very unsuccessful in mating. If electronically synthesized songs are substituted for the normal ones, the males mating success is restored (Bennet-Clark and Ewing, 1967). However, if the inter-pulse-intervals of the pulse songs are either one half or twice the normal value, the artificial songs are ineffective. Changing the duration of the pulses, which effectively means their carrier frequency, does not, however, reduce their stimulating value (Bennet-Clark and Ewing 1969). This suggests once again that the temporal features of song are important for species recognition.

The majority of *Drosophila* species, including most of the *melanogaster* subgroup, perform two distinct courtship songs. In *D. melanogaster* itself the song consists of a pulse train of single-cycle pulses (pulse song) and a low-frequency tone burst (sine song) (Figure 8.5). By looking at the effects of synthesized song on male success, von Schilcher (1976b) was able to demonstrate that the main function of sine song was to sexually stimulate females. Females which had been subjected to a period of sine song prior to being introduced to males copulated more readily. Pulse song, on the other hand, was not 'remembered' in the same way by females but acted in a trigger-like manner suggesting that it functions as a species identifier.

A rather similar conclusion was reached by Ikeda *et al.* (1980) working with *D. mercatorum*. The females of this species display an acceptance response of wing spreading which is characteristic and unmistakable. They will not copulate until they perform this, and males do not normally attempt to mate until it is performed. The posture can therefore be used as a behavioural assay of female sexual receptivity. *D. mercatorum* males have two songs, 'A' and 'B' song (Figure 8.3). Ikeda and his colleagues used a double cell of concentric construction, which in the inner compartment held courting males and females visually isolated from females in the outer one. These isolated females performed the acceptance posture even although they were not directly courted and did so only in response to 'B' song. This again indicates different roles for the two courtship song types with 'B' song in *D. mercatorum* being the equivalent of *D. melanogaster* pulse song.

Species-specific vibratory signals are performed by male web-building spiders of the genus *Amaurobius* which drum the webs of females with their pedipalps and abdomens. The initial barrier to interspecific courtship is due to pheromones secreted by females along with the web silk, and males normally drum only on the webs of conspecifics. One species, *A. ferox*, does not, however, appear to impregnate its web with a pheromone. Males of this species perform the most complex vibratory signals of all the species studied (Kraaft 1978). This is possibly because of the additional weight given to vibrational signals as an isolating mechanism in the absence of species-specific chemical cues.

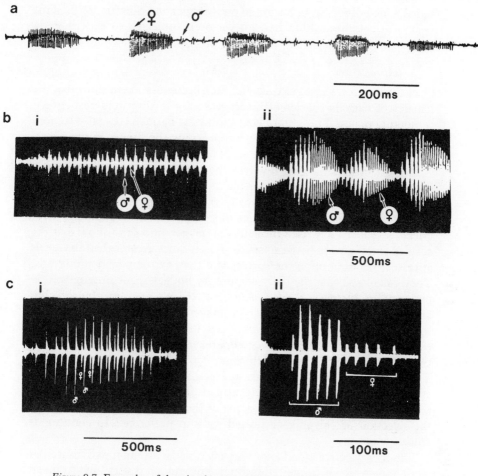

Figure 8.7. Examples of duetting between males and females. a, *Zaprionus tuberculatus*, Female sound bursts against a background of male pulse song; b, Vibratory signals of the lacewings, i, *Chrysopa carnea* and ii, *C. downesi*; c, Drumming signals of stoneflies. i, *I. quinquepunctata* and ii, *I. fulva*. (From Bennet-Clark *et al.* 1980, Henry 1979, Szczytko and Stewart 1979).

One of the several possible functions of duetting between the sexes is to ensure that they belong to the same species. As mentioned above, this may be particularly important in restricted environments where inter-specific courtships are common. However, it is a fairly widespread phenomenon, having being described not only in drosophilids and Hemiptera, but also in fiddler crabs, lacewings (Chrysopidae), and stoneflies (Plecoptera) (Figure 8.7) (von Hagen 1984, Henry 1980, Stewart, Szczytko, Stark and Zeigler 1982). In some stoneflies the signals are three-way. The male produces a characteristic pattern of pulses to which the female replies, and the male then responds with a signal which differs from his initial one (Zeigler and Stewart 1977). This system of communication has not yet been subjected to experimental analysis to investigate the function of this complex behaviour.

Sexual selection

Over the past few years evidence has been accumulating to show that the females of various species of insect use song to discriminate between competing males. This ability could form the basis for sexual selection. If so, the songs must provide cues concerning the fitness of the males, and there are several possible sources of this information.

In most insects, singing involves the use of large blocks of muscles which have evolved primarily for flight or locomotion. Song production is therefore likely to be energetically expensive. Prestwich and Walker (1981) have studied the energetics of song in several stridulating tree-cricket species. They used measures of oxygen uptake to show that the costs of singing were between six to sixteen times greater than the levels at rest, depending upon the species and the environmental conditions. They calculated that the daily respiratory budget spent singing was 56 per cent in *Oecanthus celerinictus* and 26 per cent in *Anurogryllus arboreus,*. These species produce prolonged trills, but chirping species, which perform fewer wing strokes per unit time, would expend less energy. This is borne out by a comparison of the trilling mole-cricket, *Gryllotalpa australis*, with a chirping species, *Teleogryllus commodus*. The latter uses only four times the resting amount of energy when stridulating, while the mole-cricket consumes thirteen times as much (Kavanagh 1987). This could be one of the factors which has favoured the selection of chirping or pulsed song in preference to continuous trills.

The energy conversion factor for song, that is, the amount of muscular work done which is converted into acoustic energy, is very low. At most, a very small percentage is converted usefully and singing is certainly a less efficient activity than flight. In the two species examined by Kavanagh, the efficiency for *T. commodus* was calculated at only 0.05 per cent, while for the mole-cricket it was 1.05 per cent. The higher figure for the latter is probably due to its habit of singing from a burrow which acts as an acoustic transducer. Males which sing louder and longer are therefore in some senses likely to be fitter individuals and might be selected by females.

The extent to which an animal sings or, indeed, if it sings at all, may depend upon whether it occupies a territory and on the quality of the resource that it holds. There are numerous studies on a wide variety of animals, including arthropods, which show conclusively that territory holders are at an advantage in obtaining mates (Huntingford and Turner 1987). It is worth noting that females, in selecting a territorial male, may be indirectly choosing him partly on the basis of his singing performance, as song may be involved in setting up and defending a territory. Even among aggregating species such as those which form leks, song may be involved in establishing dominance and this will ultimately affect male success.

Examples of sexual selection are known to exist in the extreme case where a choice is made between singing and non-singing individuals. This comparison is only worth making where such a large variation in acoustic behaviour is actually found in nature. One example which is particularly informative because it evaluates lifetime success is in the desert grasshopper, *Ligurotettix coquilletti*. Males vary both from day to day and during their lifetimes in the amount of time they spend singing and some males have periods during which they are totally silent. The males tend to aggregate on creosote bushes (*Larrea tridentata*) on which they feed. It is not clear whether these aggregations are active or passive, but the latter is likely as the grasshoppers congregate on those plants which contain low titres of the poisonous compound, nor-dihydroguaiaretic acid. The amount of stridulation by males is related to their social status, with the dominant males singing most. These males have the greatest mating success when measured either on an instantaneous or a lifetime basis (Greenfield and Shelly 1985).

As might be predicted, the loudness and/or amount of song produced by males have been shown to be correlated with mating success. Both of these parameters of song are often associated with body size, a factor which has often been implicated in mate choice, with females almost invariably preferring larger males. Thus, for example, large males of *Gryllus bimaculatus* have been shown to enjoy greater lifetime mating success than small males. The songs of large males are more intense and also have higher pulse and chirp repetition rates. This effectively means that they produce more song per unit time. Females appear to sample the calling songs of males within their aggregations before choosing a mate. Song loudness is not a reliable criterion of male size on its own because of attenuation. Playback experiments in which the intensity of song was controlled demonstrated that females used pulse and chirp repetition rates to discriminate between males (Simmons 1988).

Female field crickets, *Gryllus integer*, also preferentially orient towards males which produce longer bouts and therefore more song (Hedrick 1986). There is considerable natural variation for song bout length within natural populations and this has been shown to have a heritable basis (Hedrick 1988). However, in this species of cricket no positive correlation was found between body size and bout length or between age and bout length. This contrasts with what was found in the related species, *G. pennsylvanicus* and *G. veletis*, where females chose males that were older on the basis of their songs (Zuk 1987). It is not known at present which characteristics of the songs the females are capable of perceiving, and which also change with age, but older males may be the ones which have established stable territories and which therefore sing more persistently. Crankshaw (1979) also showed that *Acheta domesticus* females were preferentially attracted to the songs of dominant males. Here too, the feature of the songs which provides cues concerning social status has not been identified.

The precise basis for mate selection in these closely related gryllid species is therefore not at present clear. Dominance status, body size, age, song intensity and persistence of singing may all be correlated to some extent one with another. It is likely that females can make use of a combination of acoustic cues when choosing a mate and the relative importance of these may turn out to be different in different species.

Examples of females preferring louder songs have also been described in mole-crickets (Forrest 1983) and the bush-cricket, *Conocephalus nigropleurum* (Gwynne 1982). In the latter, the louder songs are produced by larger males. This is clearly of advantage to females as males produce spermatophores which the females eat following copulation. These have been shown to provide nutrients for egg production and larger males produce larger spermatophores.

It is not clear if song intensity per se is being assessed in *C. nigropleurum*. In another bush-cricket, *Tettigonia cantans*, Latimer and Sippel (1987) also have shown that females prefer louder songs. However, a more important factor is the relative intensity of the high- and low-frequency components of the song. Females selectively choose loud songs which contain a dominant high-frequency component. As the high frequencies attenuate rapidly with distance, particularly in dense vegetation, females are orienting not just to males which sing louder but to those which are nearby. Unfortunately no clear correlation has been established in this species between males which sing loudly and any particular physical attribute such as body size which might indicate greater fitness. Females are known to move around for several hours sampling the songs of different males before making a choice of mate, and the final decision could be based on several factors including song quality.

Rather contradictory results have been obtained from observations made by Gwynne and Bailey (1988) on an Australian zaprochiline bush-cricket. Laboratory experiments indicated that females chose males with higher-frequency calls. However, large males have larger spermatophores but lower-frequency calling songs and might therefore be expected to be preferred as mates. An examination of copulating pairs in the wild however, failed to show a positive correlation between body size and success. Gwynne and Bailey suggest that, as this species is a pollen eater, the high protein diet renders the nutrient content of the spermatophore relatively unimportant, and females tend to mate with males that are nearby.

The Diptera also provide examples of sexual selection related to song differences and to body size. Females of the Carribean Fruit Fly, *Anastrepha suspensa*, mate preferentially with large males. The songs of such males are up to 10 dB louder than those of small males and they sing for a greater proportion of the time (Burk and Webb 1983).

In *Drosophila melanogaster* also, large males are at an advantage in mating. It is difficult here, as in many of the other examples of supposed female choice, to be quite certain that the females do indeed exercise choice of mate. Large male *D. melanogaster* are better able to keep up with moving females during courtship. One consequence of this is that they spend more time singing than small males do. The extreme attenuation of the near field acoustic signal makes it unlikely that the females can hear all of the songs sung by males. As large males keep closer to females during courtship than small males, females will hear a greater proportion of their song. In addition, large males do sing more loudly than small males (Partridge, Ewing and Chandler 1987). Here again the evidence implicates the quantity of acoustic stimulation rather than any aspect of its quality in mate choice.

Although the possibility cannot be excluded, sound intensity has not been shown to be involved in mate choice in the grasshopper, *Chorthippus brunneus*. However, Butlin, Hewitt and Webb (1985) have demonstrated stabilizing selection for syllable length in this species, and females mate preferentially with males whose songs have syllable lengths near the species mean of 18.7 ms. One might expect this type of discrimination to occur where the acoustic signal is important as a species recognition or isolating mechanism. Butlin and Hewitt (1987) have further showed that the tendency of male stridulation to contain regularly spaced triplets of syllables was correlated with sexual success.

Female insects can and do therefore discriminate between males on the basis of their songs. At present there are too few detailed examples to allow us to come to any general conclusions concerning the parameters females may use and how these are related to fitness. This kind of information is becoming available in, for example, amphibia and birds, in which the problems have been addressed for a longer period than in arthropods (Searcy and Andersson 1986).

Aggressive song

As described in Chapter 7, calling songs are not necessarily exclusively concerned with attracting conspecific females. They may also play a role in establishing dominance, maintaining a territory and in overt aggression. Observations on the behaviour of males and a limited number of experiments have established the aggressive function of calling song in a number of insects.

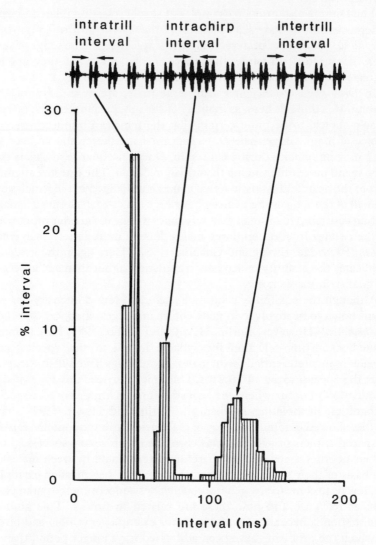

Figure 8.8. Complex calling song of the cricket, *Teleogryllus oceanicus*. The relationship between chirps, trills and the intervals between them is illustrated. The song probably serves the functions of attracting conspecific females while at the same time repelling other males. (After Bentley and Hoy 1974).

If male bush-crickets, *Pterophylla camellifolia*, call in close proximity to one another, one eventually falls silent and moves away (Shaw 1974). In chorusing *Neoconocephalus nebrascensis*, it is the male which consistently sings second, the follower male, which loses an aggressive encounter and flies or crawls away (Meixner and Shaw 1979). In the majority of such cases the songs alone are sufficient to effect inter-male spacing and overt aggression is usually avoided. Some species of bush-cricket, such as *Tettigonia cantans*, have been observed to fight and to stridulate at the same time. In this species the male which possesses the song with the lower carrier frequency usually wins the encounter and becomes dominant (Latimer and Schatral 1986). The causality here is not clear and it is probably not the low frequency itself with confers the advantage but rather that it is correlated with some as yet unknown trait.

The importance of persistent singing by males is demonstrated by some field experiments carried out on the Western Australian bush-cricket, *Mygalopsis marki* (Dadour and Bailey 1985). The males of this species space out and tend to take up singing stations on new fronds arising from the crowns of 'grass trees', *Xanthorrhoea preissii*. If a recording of the calling song is played loudly to one of these territorial males, he tends to move down the bush and finally falls silent.

The calling songs of many species of insect, such as the tettigoniids discussed above, tend to have a relatively simple temporal structure. However, this is not necessarily always the case. For example, *Teleogryllus oceanicus* sings a complex song which is composed of a mixture of chirps and trills (Figure 8.8) and Pollack and Hoy (1981), by looking at their phonotactic responses, have shown that females respond only to the chirp component. If given the choice between synthesized songs consisting entirely of chirps, and songs which resemble the normal calling songs, they prefer the former. The trill component appears to be of significance only to other males and mediates male spacing (Pollack 1982).

Similarly, some of the water bugs (*Cenocorixa* species) have a bipartite calling song which the males sing either in its entirety or they sing only the first part. Females respond to the complete song by reply stridulation while other males respond solely to the initial portion of the song (Jansson 1973). These two examples make an interesting comparison with observations on the tree-frog, *Eleutherodactylus coqui* (Nairns and Capranica 1976). The specific name is derived from its two-note call, *co-qui*. Females respond phonotactically to the *qui* portion of the call while the *co* note elicits male aggression.

Thus, in the majority of cases there are separate songs for calling and aggression, but there are also songs which are comprised of two discrete portions each with a different function and, in some cases, as in many of the Tettigoniidae, a uniform song serving both aggression and court-ship. One should note nevertheless that even in those crickets and grasshoppers which have aggressive or rivalry songs found in the context of male–male aggression, the calling songs may be additionally involved in male spacing.

Even among closely related species there appears to be considerable variation in the occurrence and form of aggressive songs. Some of the *Cenocoryxa* and *Sigara* species have distinct aggressive songs, while in others they are absent and no ecological or behavioural explanation has been put forward to account for this. Alexander (1961), who has looked at the aggressive behaviour of many North American species of ground crickets, has noted that the aggressive songs are derived from calling songs and that they have diverged to varying extents in different species. One might predict that the most aggressive species would have the most distinct rivalry songs, but this does not appear to be so.

Fights resulting in physical damage are rare in animals and are usually avoided by means of threat displays which are often visual or acoustic. Some insects have been observed to fight, and aggressive songs can act as a threat while asserting or maintaining dominance between individuals. In *Acheta domesticus* and *Gryllus pennsylvanicus* for example, it has been shown that the aggressive songs stabilize the dominance hierarchy. Subordinate males which are deafened by destruction of the auditory tympani become much more aggressive and will then fight dominant animals. The rivalry songs of dominants normally inhibit such overt aggressive behaviour (Phillips and Konishi 1973).

Also using crickets, Alexander (1961) was able to 'defeat' dominants by inducing them to fight a model. He provided artificial stimulation by mimicking the antennal lashing which is a normal component of aggressive behaviour. These defeats were much more easily accomplished if the antennal lashing was accompanied by playback of aggressive song.

In the gregarious cricket, *Amphiacusta maya*, males sing during courtship and aggression, the song being the same in both contexts. Aggressive encounters result in the setting up of a dominance hierarchy in which dominant males are more likely to mate than subordinates. Males which were silenced by waxing their stridulatory apparatus lost fights with males of equal social status, demonstrating that song is important in deciding the outcome of aggressive encounters and in establishing rank. Silenced males, even if dominant, were less successful in copulating. This was not due to an inability to provide adequate sexual stimulation, but rather because during courtship, song deters interruption by other males (Boake and Capranica 1982).

Songs which could be classified as aggressive are also produced by sexually unreceptive females of some species of insect. Wing flicks which produce loud pulses of sound are performed by unreceptive *Drosophila* females in response to male courtship, although there is no convincing evidence that this behaviour actually inhibits male courtship (Ewing 1983). Similar female rejection signals have been described in water beetles of the genus *Tropisternus* (Ryker 1976) and the sexton beetle, *Geotrupes* (Hemming and Jansson 1980). It is important that females

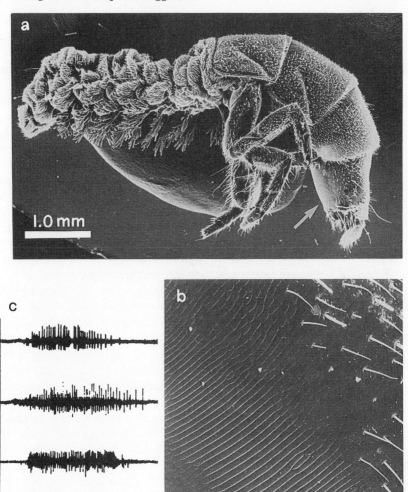

Figure 8.9. Stridulation in a caddis-fly larva, *Hydropsyche* species. a,
Scanning electron micrograph of *Hydropsyche* larva showing the position of
the files on the ventral side of the head (arrowed) and, b, a higher power
scanning micrograph of the file; c, Sound pulses of four species. i, *H.
angustipennis*; ii, *H. siltalia*; iii, *H. nevae*; iv, *H. pellicidula*. (c From Jansson
and Vuoristo 1979).

reject heterospecific courtships as well as those of their own species in situations where several related species are congregated. This probably accounts for the observation that both in *Drosophila* and *Tropisternus* the rejection signals are similar in the different species, which is in contrast to the species-specific male courtship songs.

All of the foregoing examples of aggressive song are associated with aspects of reproductive behaviour. However, there is one example of resource holding by an insect larva which involves acoustic signalling. Caddis-fly larvae of the family Hydropsychidae live on stream bottoms where they inhabit retreats made from silk attached to stones. The larvae often relinquish their homes and may attempt to evict other individuals living in more preferred sites. Sometimes fierce fights take place for possession and these may be accompanied by stridulation. This is achieved by scrapers on the fore legs being drawn across a series of ridges on the sides of the head. (Figure 8.9). These signals are probably perceived as vibrations, possibly through the substrate or the feeding webs. When the larvae are clearly matched in size, the occupier normally wins and retains possession. However, the proportion of fights won by defenders decreases with increasing discrepancy in size between the combatants, and sufficiently large attackers can evict smaller larvae from their retreats. However, if the defender stridulates during the contest then it invariably wins, regardless of the attacker's size. It is clearly advantageous for an attacker to evict a resident before it has the opportunity to stridulate, and many successful take-overs are indeed preceded by very short fights (Jansson and Vuoristo 1979). As with female rejection songs, larval aggressive stridulation is not species-specific and interspecific fights could well occur.

9. Acoustic Communication in Social Arthropods

Some of the most interesting arthropod behaviour is found in the social insects. With the need for complex communication to mediate social interactions one might expect that such insects would make extensive use of acoustic cues, particularly as many of them live in the dark where visual communication is not possible. However, this does not appear to be the case, and relatively few examples of acoustic signalling have been described from social and subsocial arthropods. One reason for this is that pheromones are a more appropriate means of regulating the relatively long-lasting behavioural changes that take place in beehives and ant nests. Acoustic stimuli are mainly used where immediate messages need to be passed.

SUBSOCIAL INSECTS

There are examples of subsocial behaviour in many insect orders and in other arthropods such as spiders, crustacea, mites and millipedes, but very few cases indeed of acoustic behaviour have been described from them. A simple example, however, is found in the larvae of the colonial sawfly, *Hemichroa crocea*, which feed communally. There are several advantages in feeding together, such as a reduction in the level of predation and the greater probability of one member of a colony penetrating tough plant tissue, enabling the remainder to feed also. When local food supply is becoming depleted, *H. crocea* larvae stridulate by dragging the tips of their abdomens along the leaf surface. This acts as a substrate-borne vibrational signal which induces the larvae to migrate to a fresh leaf, thus maintaining the coherence of the colony (Hograefe 1984).

Beetles from many different families produce stridulatory or percussive sounds. While most of these are presumed to be warning or disturbance signals, in subsocial species such as the sexton (*Geotrupes* species) and wood-boring beetles (Passalidae; Scotylidae) they play a part in intraspecific communication (Alexander, Moore and Woodruff 1963, Rudinsky and Ryker 1976). The passalids are a mainly tropical family of beetles which live in galleries in rotten wood and have an advanced subsocial life. There is some overlap of generations and young adults help their parents to care for their pupal siblings. These beetles have a particularly rich repertoire of acoustic signals with, for example, four different types of sound being found in *Odontotaenius disjunctus*. These are produced in a large number of behavioural contexts. The larvae also stridulate, using the second and third pair of legs: the adults have an abdominal–wing mechanism, and so the two types of acoustic behaviour have evolved independently. It is tempting to suppose that the complexity of the acoustic communication is a consequence of the insects subsocial behaviour. However, the contexts in which these signals occur, male and female aggression, courtship and copulation are not peculiar to social life (Schuster 1983). The subsocial cricket, *Anurogryllus arboreus*, also has a large acoustic repertoire of six distinct signals, but again none of these are observed in contexts different from those in non-social crickets (Alexander 1962).

SOCIAL INSECTS: THE ANTS

Many species of ant produce sound by means of an abdominal–petiolar stridulatory mechanism (Markl 1973). Possible functions of this beha-viour have been investigated in harvester ants, *Pogonomyrmex* species by Markl, Holldobler and Holldobler (1977). Although several closely related species live allopatrically in the south-western United States and there is the possibility of interbreeding between them, no evidence was found for acoustic isolating mechanisms. Mating occurs at leks where the new queens copulate several times with different males. The females stridulate loudly when the males attempt to prevent them from leaving the lek after having mated and these signals have therefore been termed 'female liberation signals'. The other context where stridulation is observed is during fights, when young queens compete for founding burrows. Although no influence on the eventual outcome of such fights has been attributed to stridulation, in view of other work on aggressive sound it seems likely that some such effect must exist.

Carpenter ants, *Camponotus herculeanum*, produce drumming signals which are widely transmitted through the wood in which they construct their nests (see Chapter 2). These signals have been shown to affect the behaviour of ants both within the nest itself and on the surrounding tree-trunk. Two different responses have been observed. The majority of ants, particularly the slow-moving individuals, cease movement and freeze. For the ants outside the nest this may constitute an anti-predator device by making them less obvious to insectivorous birds. On the other hand, for faster-moving workers, the vibrations have the opposite effect, and these ants run around frantically with jaws agape (Fuchs 1976). These signals do therefore play a part in the social behaviour of carpen-ter ants by co-ordinating defensive behaviour. A similar function has been ascribed to the percussive signals produced by termites (Howse 1964).

Social insects: the bees

Among the social insects it is the bees that make most use of vibrational and acoustic signals, both in regulating hive activity and in food gathering. During the performance of the well-known waggle dance of the Honey Bee, *Apis mellifora*, workers produce a 280 Hz acoustic signal by vibrating their wings. The function of the dance is to recruit foragers to a food source and to provide information concerning the direction and distance of the source from the hive. The orientation of the dance with respect to gravity indicates the former. Distance is correlated with the duration of the waggle run and therefore also with the length of the acoustic signal. Follower bees stay very close to workers while they dance and could obtain this information from either tactile or acoustic cues.

Conclusive evidence on the relative importance of these two different sources of information is lacking at present but it is fairly clear that an acoustic component is involved. Esch (1963) noted that some dances occur without the production of sound but are otherwise normal and that they fail to recruit foragers. A comparative study of four different *Apis* species revealed that two of them, *A. dorsata* and *A. florea*, which build their nests in the open, do not produce sound during the waggle dance. By contrast, *A. cerana* which, like *A. mellifera*, lives in cavities, does produce sound. This suggests that the acoustic signals have evolved as an adaptation for communication in the dark (Towne 1985).

Some of the social bees, such as bumble bees (*Bombus* species), have not been observed to use sound for communicating information concerning food sources. The stingless bees of South America, however, do use sound in this way, although in a much less advanced form than the Honey Bee. Their behaviour indicates the way that the waggle dance may have evolved. Thus, workers of *Melipona* species produce sounds on returning to the nest after a successful foraging flight and, as in *A. mellifora*, the duration of the sounds is correlated with the distance of the food source. However, the acoustic information alone is apparently inadequate to guide workers to the food. In addition to providing acoustic stimuli, foragers lead naïve workers in the direction of the source by means of repeated zig-zag flights.

Trigonia species show even less developed acoustic signalling. The sounds produced by forager bees contain no information on the distance of the hive from the food, but merely excite the workers. They only attempt to find the source if food is offered to them by returning foragers (Esch 1967). These signals therefore prime the appropriate behavioural response which is only fully released by additional stimuli. A similar situation is found in ants of the genus *Novomessar* where stridulatory signals facilitate food retrieval by workers (Markl and Holldobler 1978).

In addition to acoustic signals, Honey Bees (*A. mellifera*) also produce a vibrational signal whose carrier frequency is 320 Hz. The effect of substrate vibrations on Honey Bees is to make them immobile. Unlike the response of carpenter ants, this is not concerned with defence of the colony. During the waggle dance, follower workers produce these vibrational signals which induce the dancing bees to stop momentarily and regurgitate some food. This has therefore been called a 'begging' signal (Esch 1963, Michelsen, Kirchner and Lindauer 1986).

The remaining socially significant acoustic signal of honey bees is the so-called 'piping' which is performed by the queens. The function and mechanism whereby these sounds are produced have intrigued apiarists for centuries. Piping comprises two distinct sounds – 'tooting' which is done by newly emerged queen bees and 'quacking' by queens before they emerge from their pupal cells (Figure 2.8). While beekeepers have long been aware of these sounds and have known that they are associated with swarming, their true function has only recently started to become clear.

Tooting is performed by a newly emerged queen and is heard about 7 to 10 days after the departure of a prime swarm, that is, the initial swarm of the year which contains the old queen. Experiments by Bruinsma, Kruijt and van Dusseldorp (1981) in which they made use of play backs of tooting and quacking, demonstrated that the effect of the former is to delay by two or three days the emergence of queens that are still in their cells. It is not entirely clear how this is brought about, but it is in part mediated by workers. They prevent the queens from hatching by resealing the cells as the queens try to cut their way out. In addition, workers perform movements termed dorso-ventral abdominal vibrations (D-VAV) on top of the cells, and these probably produce tactile stimuli which inhibit the queens from emerging. It has also been observed that young queens are subjected directly to D-VAV and the effect of this is to induce queens to toot. Queen emergence and swarming are therefore under the control of the workers rather than the queen.

The delay in emergence of the queens from the cells due to tooting allows a second swarm to leave the hive with the young queen. This process may be repeated several times if the colony is sufficiently strong or if it has become overcrowded. If the young queen does not toot, then the subsequent queens are killed, either just before, or soon after emergence.

The role of quacking is much less clear. It is almost invariably done by unemerged queens in response to tooting by the resident queen. It is presumably a reply which inhibits the aggression towards her until such time as it is possible for her to emerge in safety.

10. Genetics and Evolution of Song

Behaviour is the outcome of genetic, developmental and environmental influences and interactions which are furthest removed from the site of primary gene action. Behavioural traits are themselves nevertheless subject to selective forces and evolutionary pressures in the same way as biochemical, morphological or physiological characters.

Arthropod acoustic behaviour provides an almost ideal system for studying the inheritance of behaviour. One problem for behaviour genetics has always been the difficulty of specifying appropriate behavioural characters which can be scored. The sounds produced by most arthropods are stereotyped and easily characterized. In this respect they are more like morphological than behavioural characters. Further, as most sounds are involved in sexual behaviour they clearly have considerable adaptive significance and it is therefore important to understand their underlying genetic control and how they have evolved. It must be noted however, that while considerable work has been done on the genetics of sound production, very little has been carried out on the reception of sound.

Much of our understanding of the genetics of sound production comes from work on that most versatile of experimental animals, *Drosophila*. Most species within the genus possess a complex acoustic repertoire in addition to their well-known advantages for genetic research, of short generation time, a well-mapped genome and ease of maintenance. A limited amount of research has also been carried out with crickets, grasshoppers and water bugs.

GENETIC CONTROL OF SONG

There are a number of experiments that demonstrate genetic variability for song parameters and that these can be altered by selection. For example Ikeda, Takabatake and Sawada (1980) have shown that the courtship songs of *D. mercatorum* are strain dependent. The amount of song produced during courtship has also been shown to be important in a variety of arthropods for reasons discussed in Chapter 8. The alternative strategies of having calling and silent satellite males in the cricket, *Gryllus intiger*, is under genetic control and may be an example of a balanced genetic polymorphism. Artificial selection for change in the amount of calling by males resulted in a rapid response over very few generations for both increased and decreased levels of calling when compared with the unselected population. This suggests that rather few alleles are involved in controlling the trait (Cade 1981b).

There is both indirect and direct evidence that the amount of courtship song performed is also under genetic control in *Drosophila* species. In lines of *D. melanogaster* which had been selected for large and small body size the proportion of song in their courtships was reduced by 16 per cent in the former and increased by 36 per cent in the latter if compared with control flies. An indirect effect of change in body size had been to alter the mating success of the small and large males. This had resulted in selection for compensating changes in their courtship behaviour which was manifest in the amount of song produced (Ewing 1961).

McDonald and Crossley (1982) were successful in selecting directly for changed proportion of song in *D. melanogaster* courtship. Interestingly, selection for this character changed other behaviours such as levels of locomotor activity and the inter-pulse-intervals within pulse song. Even morphological characters such as wing area were altered, indicating rather complex pleiotropic genetic effects. The converse experiment was done by Ikeda and Maruo (1982) who successfully carried out bi-directional selection for changed inter-pulse-interval in the song of *D. mercatorum*. They also found associated changes in the behavioural physiology and morphology of the selected lines.

These experiments demonstrate that genetic control of quantitative song characteristics exists. It is much more difficult to reveal control of qualitative aspects. One method is to screen flies for genetic mutations and this has achieved limited success. Von Schilcher (1976c) isolated a mutant allele of *D. melanogaster*, *cacophony*, which altered the nature of pulse song. *Cacophony* males produce pulses that are polycyclic in form, compared with the monocyclic pulses of wild type males, and these pulses are also twice as loud as normal. The inter-pulse-intervals are also longer than in the wild type. It is not known how this allele acts, but its features are consistent with a perturbation of the feedback loops controlling wing movement. One interesting point is that the mutant sound pulses are similar to those found in several other *Drosophila* species. This example provides a model to show how small genetic changes can have profound effects of song and lead to the kind of divergence which must have occurred during speciation.

Biometric analyses

One method that has been employed to investigate the genetic architecture of song in *Drosophila* species is the biometric analysis developed by Mather and Jinks (1982) which makes use of diallel crosses. The technique requires a series of F1 and backcrosses to be made, ideally, between isogenic genotypes. Correlations are then calculated between characters displayed by the different crosses. This allows estimates of several genetic parameters of the traits such as heritability, dominance and additivity to be made. The genetic contribution of individual chromosomes can also be deduced by using genetic markers.

This method was used by Hoikkala and Lumme (1987) to examine the inheritance of courtship song in the *D. virilis* group of species. They produced crosses both between strains within a species and also between different species which were able to hybridize. They scored several song parameters and demonstrated that the genes controlling them were, for the most part, distributed throughout the genome and acted additively. In two of the species, *D. virilis* and *D. littoralis*, however, the genes controlling inter-pulse-intervals were concentrated on the X chromosomes.

By analysing the songs of interspecific hybrids, they gained some insight into the directions of evolutionary change in song pattern. Thus, for example, the X chromosome involvement with inter-pulse-interval is associated with the separation of the *virilis* group of species into two distinct branches, the *virilis* and *montana* phylads. In the former, sex-linked genes which increase the length of the pulse trains and pulse durations and which decrease pulse interval are dominant. This suggests that within the species making up this phylad, directional selection has existed in favour of longer and more continuous songs and also for songs which are similar to one another.

There is much greater variability in inter-pulse-interval between species of the *montana* phylad. This has come about because the dispersion of genes controlling this trait throughout the autosomal chromosomes has provided greater possibilities for directional selection. This finding suggests that the functions of male song in the two phylads are rather different. While in the *montana* phylad the interspecific differences in temporal pattern are sufficiently large for the songs to act as isolating mechanisms, the same is unlikely to be true within the *virilis* phylad (Figure 10.1).

Diallel crosses between inbred lines of *D. melanogaster* have demonstrated that both the inter-pulse-interval of pulse song and the carrier frequency of sine song (see Figure 8.5) are controlled by genes carried on the autosomes which act additively and without any strong dominance (Cowling 1980). A similar conclusion was reached by Kawanishi and Watanabe (1981) who analysed the songs of hybrids between *D. melanogaster* and *D. simulans*. The lack of dominance in a trait is often considered to indicate that it is not an important component of fitness. However, the behavioural evidence on the relevance of temporal patterning for species identification and sexual stimulation suggests otherwise. The latter view is reinforced by the observation that populations of *D. melanogaster* from many regions of the world have song inter-

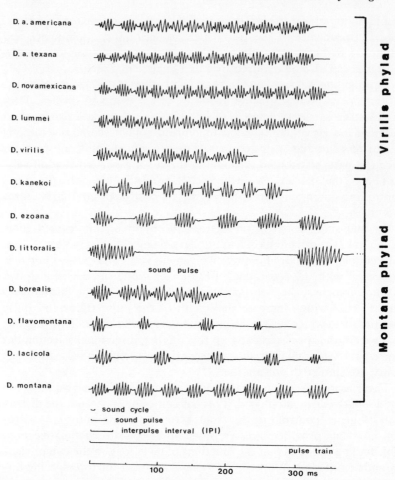

Figure 10.1. Oscillograms of the male courtship songs of the *Drosophila virilis* species group. There is little variation in inter-pulse interval between members of the *virilis* phylad when compared with the *montana* phylad. (From Hoikkala and Lumme 1987).

pulse-intervals whose mean values vary with a range of between only 31 and 35 ms. This indicates that pulse interval is a very conservative character which has been subject to strong stabilizing selection, a view strengthened by the evidence from selection experiments that genetic variation for the character exists within natural populations. By comparison, *D. simulans*, a species which makes more use of visual stimuli in its courtship, has much more variable songs. Mean values of inter-pulse-intervals in different populations of this species vary between 42 and 57 ms (Kawanishi and Watanabe 1980).

Song in interspecific hybrids

Interspecific crosses have been made with the aim of investigating the inheritance of song in a number of different insect species. In *Drosophila* they have been made between *D. melanogaster* and *D. simulans* (von Schilcher and Manning 1975, Kawanishi and Watanabe 1981), *D. virilus* and *D. lummei* (Hoikkala and Lumme 1984) and *D. persimilus* and *D. pseudoobscura* (Ewing 1969); in crickets between *Teleogryllus oceanicus* and *T. commodus* and between *Gryllus argentinus* and *G. peruviensis* (Leroy 1964, 1965, Hoy and Paul 1973); in grasshoppers between *Chorthippus* species (Perdeck 1957, von Helversen and von Helversen 1975) and in the water bugs *Arctocorisa carinata* and *A. germari* (Jansson 1979). One result common to all of these experiments is that many of the song parameters in the hybrids were intermediate between the parental species. This was particularly true of temporal characters such as chirp and pulse repetition rates, and suggests that these are multifactorially controlled with little directional dominance. In examples where the individual chromosomes bore markers, the genes were seen to be distributed throughout the genome. This reinforces the conclusions derived from the biometrical analyses described above concerning the additive, polygenic nature of genetic control.

One interesting result which emerged from some of these crosses was evidence of sex-linkage in certain acoustic parameters. For example, *D. persimilus* and *D. pseudoobscura* males each perform two different song types (Figure 8.4). In the F1s and backcrosses, the inter-pulse-intervals were intermediate between those of the parental types indicating polygenic control. Which of the two types of song was performed depended, however, upon the maternal X-chromosome. Sex-linkage for song characters has also been found in *Chorthippus parallelus* (Butlin and Hewitt 1988) and *Teleogryllus* species (Hoy and Paul 1973). Other sex-linked factors which have not been specifically identified but which affect mate choice have been demonstrated in *D. melanogaster* and *D. simulans* (Kawanishi and Watanabe 1981, Wood and Ringo 1980) and *D. arizonensis* and *D. mojavensis* (Zouros 1981).

One consequence of sex-linkage is the potential for more rapid evolutionary change than is possible for autosomally controlled characters. As the male is the heterogametic sex in insects, any genotypic change affecting a sex-linked character will find immediate phenotypic expression in the male and be subject to selection pressure. Fixation of advantageous, sex-linked traits in a population is therefore likely to be rapid.

Co-evolution in acoustic communication

Where acoustic signals are important for species recognition there will be strong stabilizing selection to maintain both the signal itself and the

signal recognition system at a particular value. It is difficult to envisage
how such an interdependent system could change during evolution
except extremely gradually. Any deviation from the optimum value of
the character in one sex would be most likely to be disadvantageous and
immediately selected against. However, there are many examples of
closely related species which differ one from another more in song than
in any morphological features. This is reflected in the fact that taxonom-
ists often make use of the structure of the sound-producing apparatus in
constructing keys for the identification of species (Gogala 1984).

The status of so-called cryptic species has often been determined
initially on the basis of their songs. Thus, in the intensively studied
Ensifera of the eastern United States, the number of species had become
stable by the third decade of the present century. Subsequent examination of their song patterns has resulted in the number of actual species
being increased by almost 25 per cent (Walker 1964b). Other examples
of cryptic species revealed by their songs include *Drosophila* species and
bugs of the genus *Oconopsis* (Miller, Goldstein and Patty 1975, Claridge
and Reynolds 1973). It seems likely that the detailed examination of
calling and courtship songs in many insect groups will reveal further
cryptic species.

It has often been considered that differences in song have arisen
because of the selective pressures for premating isolation. One hypothesis proposed to account for the way in which such rapid evolutionary
changes could come about is that there is a common genetic basis for
both sound production and its reception (Alexander 1962). One can
visualize a reference oscillator whose periodicity would be used in males
to set a temporal parameter such as pulse repetition rate, and also act as
a template in females against which incoming signals would be matched
(Ewing 1969). If this were so then any mutation affecting the reference
oscillator would simultaneously alter the production and recpeion of the
signal in a co-ordinated manner. In addition, if any of the genes controlling song were sex-linked, there would be immediate expression of any
genetic change in males. A weakness of this argument is that females,
being the homogametic sex, would be heterozygous at the relevant locus
and would not necessarily respond to the changed signal.

This hypothesis has been espoused by Hoy and Paul (1973) to account
for co-evolution of cricket communication, and by Doherty and
Gerhardt (1983) for tree-frogs. On the other hand, D. and O. von
Helversen (1987), working with grasshoppers, claim that, in these
insects at least, genetic coupling of reception and transmission of signals
does not occur.

There are two pieces of evidence which support the concept of genetic
coupling. F1 hybrid females from crosses between the two cricket
species, *Teleogryllus oceanicus* and *T. commodus*, are preferentially attrac-

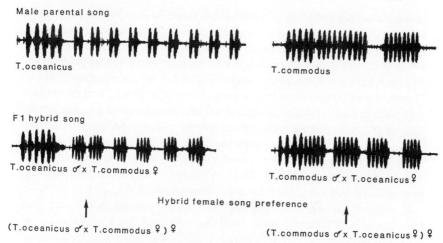

Male parental song

T.oceanicus

T.commodus

F 1 hybrid song

T.oceanicus ♂ x T.commodus ♀

T.commodus ♂ x T.oceanicus ♀

Hybrid female song preference

(T.oceanicus ♂ x T.commodus ♀) ♀

(T.commodus ♂ x T.oceanicus ♀) ♀

Figure 10.2. Phonotactic responses of hybrid female crickets. Due to sex-linkage, the songs of the two reciprocal F1 hybrids are different. Hybrid females prefer the tape-recorded songs of congenic males. (After Hoy *et al.* 1977).

ted to the songs of hybrid males when given the choice between the songs of hybrids and those of either parental species. Songs of F1 males are intermediate between those of their parents. However, because some of the song parameters are controlled by sex-linked genes, the songs of the two reciprocal hybrids differ. Hybrid females prefer the songs of their siblings to those of males of the reciprocal cross (Figure 10.2) (Hoy, Hahn and Paul 1977). This precise matching of song type and phonotactic preference is good evidence for some common mechanism in the two sexes which is under genetic control. The same phenomenon has also been established by Doherty and Gerhardt (1983) for hybrids between two tree-frog species *Hyla chrysocelis* and *H. femoralis*. Contradictory results have been obtained from hybrids between *Chorthippus* species. F1 progeny chose the songs of both of the parental species over those produced by hybrids. Von Helversen and von Helverson (1987) say that this indicates that the F1 grasshoppers possessed both of the adult templates and cite this as evidence against the idea of genetic coupling.

Another piece of evidence supporting a common genetic mechanism is the temperature coupling that has been shown to exist between song production and reception in all of the cases where it has been investigated. Some insects signal acoustically over a large range of ambient temperatures. *Cyphoderris*, for example, has been heard to stridulate from $-8°$C to $25°$C and *Gryllus* species will call over a range of 30 centigrade degrees. Change in pulse and chirp repetition rates tend to be linear over physiological temperature ranges. The Q_{10} for *Drosophila melanogaster* courtship song is about 1.6 (from Shorey 1962) and 1.7 and 1.4 for *Gryllus* chirp and pulse rates respectively (from Doherty 1985c). One would expect some type of temperature matching to occur merely because all of the physiological processes underlying sound production and reception are likely to be similarly affected by temperature change. However, the temperature dependence of song parameters varies widely between species. In the so-called 'thermometer cricket', *Oecanthus fultoni*, the relationship between chirp rate and temperature is strictly linear over the entire temperature range, while in other cricket species, pulse and chirp repetition rates tend to reach a maximum at about 25° C and drop thereafter (Walker 1962). A common mechanism underlying temperature matching would compensate for such vagaries.

Results of an ingenious experiment carried out by Bauer and von Helverson (1987) have been used as evidence against the genetic coupling hypothesis. They implanted miniature heating elements near the brain and the metathoracic ganglia of the grasshopper, *Chorthippus parallelus*, and, by selective heating, achieved a temperature differential of up to 9° C between the two regions (Figure 10.3a). By recording the females' responses to recorded male song played at different speeds so as to correspond to those of males at different temperatures, they demonstrated that the temperature of the brain was crucial for song recognition. On the other hand, the tempo of the song produced by a female depended upon the temperature of her metathorax (Figure 10.3b). This demonstrates that the neural circuitry for song recognition resides in the brain, and that for the motor output controlling stridulation in the metathorax. This is consistent with the results of neurophysiological experiments. However, while it makes the existence of a single reference oscillator most unlikely, this result does not exclude the possibility that both neuronal circuits have elements which are specified by the same genes.

The genetic coupling hypothesis remains unproven at present, but if such a mechanism does occur it is likely to have evolved independently more than once. Evidence for it exists in different groups of insects and in anurans. An analogous example has also been described in fireflies (Lampyridae, Coleoptera) which communicate by means of temporally patterned flashes of light (Carlson, Copeland, Raderman and Bulloch 1976). Nevertheless, there is no reason to expect it to be a universal phenomenon, and it may occur in some taxa and not in others.

THE EVOLUTION OF SONG

Possible evolutionary origins of arthropod acoustic communication have been described in Chapter 2. The geological record provides few clues on the evolution of behaviour such as acoustic communication and, in any case, the insect fossil record is insufficiently detailed to be of much

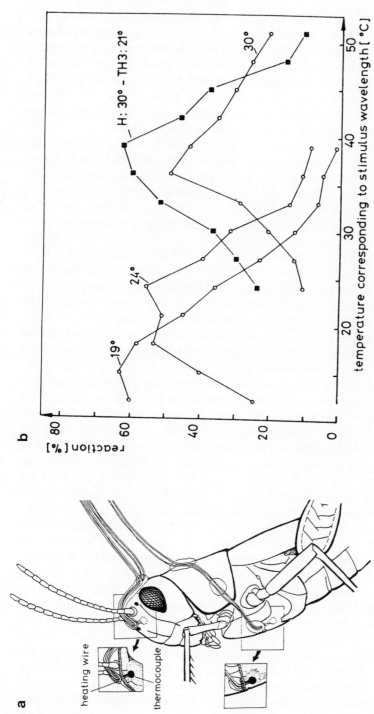

Figure 10.3. a, Experimental set-up for selectively heating the brain of *Chorthippus parallelus*. Thermocouples record the temperature near the brain and thoracic ganglia; b, Responses of a female with and without the head being heated. As the ambient temperature increases, the female responds to recorded male song played at progressively greater speeds (open circles). If the head is heated to 30 °C, and the thoracic ganglion is at 21 °C, the female's response is determined by the temperature of the head (filled squares). From Bauer and von Helversen 1987.

help. While some structures, such as those used for stridulation in gryllids, can be seen in relatively recent fossils, it is from comparative studies of extant forms that we have to infer the course of evolution.

One of the best-known sound-producing arthropod groups worldwide is the gryllids, and Alexander (1962) has identified some of the evolutionary trends that have given rise to the great variety of songs they exhibit. Primitive crickets and bush-crickets almost certainly produced trills or unmodulated pulse trains. Such songs are widely found not only in Ensifera, but also in cicadas, plant bugs and *Drosophila*. Unpatterned songs are likely to have been a primitive character in many arthropods. The motor patterns underlying continuous trilling in crickets closely resemble those of flight. The neuronal control of both are so alike as to make it almost certain that stridulation in Ensifera is derived from either flight or flight intention movements. Interestingly, stridulation in grasshoppers, which involves leg movements, is similarly related to flight and not, as might be expected, to walking. The muscles involved are bifunctional and move both the wings and the legs. The motor co-ordination of the same muscles is different in the two contexts (Elsner 1983).

The repetitive, cyclical pattern of wing movements during flight pre-adapts the system for acoustic communication. In crickets, the flight motor is driven by a neural oscillator which has a periodicity of around 30 Hz. The syllable repetition rate of song is also 30 Hz, suggesting a common timing mechanism for the two behaviours (Kutsch 1969). Similarly, in the grasshopper, *Chorthippus biguttulus*, the complex pattern of leg movements during stridulation becomes much simplified if peripheral feedback from the legs is eliminated. In this circumstance the motor output during putative stridulation occurs at the flight frequency of 50 Hz (Elsner 1983). This again indicates the existence of neuronal circuits common to the two behaviours. In *Drosophila* also, sine song is produced by wing movements which are homologous to low-amplitude flight.

A subsequent stage in the evolution of song will have been the partitioning of long pulse trains or trills into chirps. The selection pressures for this step will have been threefold. First, as discussed in Chapter 8, singing is energetically expensive and the less song produced per unit time, the less energy expended. Second, intermittent songs will be less likely to be intercepted by possible predators and parasites and third, the grouping of pulses into chirps of various lengths will have facilitated the evolution of more complex coding. Additional information might then be conveyed by songs which could become species- and context-specific.

In those arthropod groups whose present-day representatives possess specialized sound-producing organs, such as the stridulatory mechanisms of crickets and grasshoppers, the initial sounds must have been quiet and used for communication over a short distance. They must therefore have been courtship rather than calling songs. Alexander's comparative study of cricket song suggests that the direction of evolution has indeed been in this direction, with calling and aggressive songs having been derived from courtship.

The overall evolutionary trends from simple to more complex patterning of song, accompanied by a divergence of song types within a species to convey different messages, are easily appreciated. The selective pressures that have brought about the subsequent radiation of song pattern are much less clear however. One point of conflict with regard to the evolution of courtship behaviour has been the extent to which signals have been shaped by sexual selection or selection for premating isolation or by incidental influences. Until fairly recently almost all of the differences in calling and courtship songs between related species have been attributed to the need to maintain and reinforce sexual isolation (Mayr 1972). Courtship and mating of heterospecifics is wasteful both of reproductive effort and of gametes. Interspecific hybrids, if they occur, are likely to be sterile or of reduced viability. Selection would therefore be expected to favour factors leading to premating isolation.

Some workers however, have criticised the concept of selection acting directly on characters which are involved in sexual isolation during the course of speciation (Paterson 1985, Butlin 1987). There is no disagreement on the role of song in maintaining premating isolation in arthropods and other animals. The point of contention is the process whereby differences in acoustic signals have arisen.

Thus Toms (1985), working with South African tree-crickets (Oecanthinae), suggests that differences arose for the most part because of selective pressures unrelated to the signal functions of song. In tree-crickets the proportion and mass of the elytra differ considerably between species. Furthermore, experimental mutilation or loading of the elytra with small weights significantly changed the parameters of the songs. Toms argues that the changes in body dimensions have occurred due to radiation into different microhabitats with diverse ecological requirements. Changes in song, which may well be important in species recognition, will nevertheless be secondary and not themselves selected for.

While a process such as that outlined above will certainly have occurred, it is unlikely to have been a major source of interspecific variation. There is little positive evidence for a correlation between physical features and the acoustic parameters important in communication. For example, Keuper, Weidemann, Kalmring and Kaminski (1988) compared the songs of seven European tettigoniids selected because they exhibited a large range of physical characteristics such as size, weight and wing area. They could find no evidence of any association between identifiable acoustic and physical parameters, nor were they able to

correlate the song types of the different species with particular habitats
in which they lived.

The genus *Drosophila*, which has relatively little interspecific morphological variation, provides examples of very diverse patterns of
courtship behaviour. Closely related species differ widely in the use they
make of visual, pheromonal, acoustic and vibrational stimuli during
courtship (Ewing 1983, Hoy *et al.* 1988).

Conversely there are many examples, some of which have been discussed earlier in this chapter, of cryptic, sympatric species which differ
primarily in their song patterns. By contrast, many allopatric, but less
closely related species, have very similar songs. This seems to demonstrate that species identification signals are usually conservative except
where the possibility of interspecific mating exists.

Sexual selection has been invoked as a major agent in bringing about
the large differences in the acoustic repertoires of related species (West-
Eberhard 1984). Inter-male competition allied to female choice might
result in selection for males which sing louder, longer or more ex-
travagantly. This in turn would lead to a rapid change in some of these
parameters. It is difficult to determine whether signal divergence has
occurred due to sexual selection or to selection for premating isolation.
One prediction which can be made is that if the latter had operated one
would find examples of character displacement. That is, where a zone of
overlap exists between two partially distinct populations, those factors
which reinforce premating isolation should be strongly selected for.
They are likely therefore to be more different in areas of population
overlap than elsewhere. The paucity of convincing examples of character
displacement of song characters has been used as an argument against
selection for sexual isolation having occurred. However, it should be
kept in mind that character displacement would only be expected to exist
for a brief period of evolutionary time. If the populations merged, there
would be strong selection for one or other of the song patterns while, if
they remained isolated, any change in acoustic signal which reinforced
isolation between the populations would spread rapidly.

There are nevertheless some examples which can be attributed to
character displacement. One such is provided by the sibling fiddler crab
species *Uca mordax* and *U. burgersi*. Although these two species are
morphologically very similar and there is little observable difference
between their courtship displays, the vibrational components of courtship are quite distinct in both males and females. These differences are
most extreme in an area of sympatry on the island of Trinidad. The
vibration patterns of *U. burgersi* from Florida more closely resemble
those of *U. mordax* in Trinidad than those of conspecifics from the
island. What appears to have happened is that mainland *U. burgersi* and
island *U. mordax* have retained the common ancestral pattern of vibra-

tional signals, while those of Trinidad *U. burgersi* have diverged from the ancestral pattern, presumably as a result of character displacement (von Hagen 1984).

In *Drosophila* species also there are examples which can be explained most easily on the basis of character displacement. In the *melanica* group, the courtship songs of individual species are composed of a few elements drawn from a larger repertoire of distinct components. Thus each element is common to more than one, but not to all species. They are expressed in such a way that the songs of closely related species are more different than those of distantly related ones (Ewing 1980). Similarly, *D. athabasca* comprises a number of so-called 'semispecies' which can be distinguished from one another on the basis of chromosome structure and courtship song. These songs are most distinct in areas of sympatry where strong premating isolation has been demonstrated. The song patterns which are most alike are found in two of the more distantly related, allopatric semispecies (Miller *et al.* 1975).

Evidence of selection for premating isolation also comes from work by Coyne and Orr (1989) who looked at 118 pairs of *Drosophila* species of known genetic distance and geographical range. They compared sympatric and allopatric pairs and showed that in the latter, mating discrimination diverged at the same rate as post-zygotic isolation during the course of evolution. Mating discrimination can be due to a number of ecological and behavioural factors, including differences in song, while postmating isolation results from hybrid sterility or incompatibility. However, the development of premating isolation had been much more rapid than that of postmating isolation between pairs of sympatric species, and this suggests that selection for sexual isolation has occurred.

One prediction that can be made is that where sexual selection has operated on a character it will have diverged rapidly from its initial value. This would happen because females choose mates expressing extreme forms of the character. There is evidence for such 'runaway' evolutionary change affecting secondary sexual characteristics in many animals, including insects. For song, however, most of the differences that have been found between species can be attributed equally to sexual selection and selection for sexual isolation.

An example where both stabilizing selection for species recognition features of song and sexual selection may have occurred simultaneously is in the *Drosophila repleta* group of species. From cytogenetic studies of this group, the phylogenetic relationships of its constituent members are known, and it has been possible to deduce the directions of evolutionary change. A phylogenetic tree which illustrates the evolutionary history of the *repleta* group is shown in Figure 10.4. Three separate trends in the evolution of the song patterns can be identified. First, there is a tendency for closely related subgroups to possess some common, unique features.

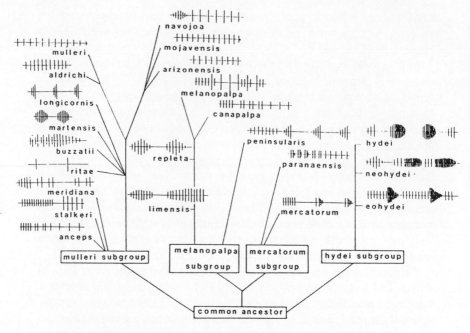

Figure 10.4. Phylogenetic tree of the *Drosophila repleta* group of species (after Wasserman 1982). The songs are shown in diagrammatic form. Oscillograms of some of the species are illustrated in Figure 8.3. (From Ewing and Miyan 1986).

This is seen best in the *hydei* subgroup whose songs alone contain long, relatively high-frequency pulses. Second, many of the species perform two distinct types of song. In all cases one of these is a simple pulse train made up of monocyclic pulses and it is produced by males early in courtship. This is a conservative character within the group found also in most of the species that have only one song type. The only difference between species is in the inter-pulse-intervals. Third, where a second type of song is found, it occurs later in courtship and tends to be complex in form and variable both within and between species.

One interpretation of these results is that the simple pulse trains are species-identifying signals which have been subject to stabilizing selection. Differences between them may have arisen due to selection for sexual isolation but direct evidence for this is lacking. The complex songs, which are performed once species identity has been established, have become diverse due to the force of sexual selection.

This distinction is almost certainly an oversimplification. An acoustic signal can be used for species recognition and at the same time affect mate choice. It is likely that sexual selection, sexual isolation and incidental effects all have had a part to play in shaping the acoustic behaviour of arthropods.

11. Bioacoustic Methods

One of the pioneers in the study of insect song was the Swede, Ossiannilsson who, just after the Second World War, described the songs of 96 different species of small plant bugs (Auchenorrhyncha, Homoptera) and also described the muscles and cuticular structures responsible for their production (Ossiannilsson 1949). Lacking the benefits of modern electronic technology, he listened to the sounds that they produced either by holding a small glass tube containing the insects to his ear or, if he wished to observe them at the same time, he used a simple device consisting of a glass tube containing the insects, around which was wound a spiral of wire. The other end of the wire was similarly wound round another tube held to his ear, making a simple telephone like the toys at one time sold to children. He described the songs, using musical notation, and either used his violin or a tuning fork to determine their pitch (Figure 11.1). He subsequently used a primitive crystal microphone which weighed 3.5 kg and a valve amplifier. In spite of the unsophisticated techniques employed, his work still constitutes a major contribution.

Since the time of Ossiannilsson and particularly over the last decade, progress in electronic scientific instrumentation and Hi-Fi has meant that the methods of making recordings of acoustic and vibrational signals are easily accessible to professional and amateur biologists alike. The development of high quality miniaturized recorders and amplifiers has facilitated field work in particular while the spread of microcomputers has opened the way to fairly simple methods of signal analysis.

As the equipment field is changing rapidly and as the author has no direct experience of much of the equipment which is in use, no attempt is made to assess specific products. Rather this chapter deals with the principles of recording and analysing arthropod sound to give some idea of appropriate methods. Most scientific papers on the subject give details of the apparatus used, and it is often apparent that what is easily available, rather than what is ideal, is used.

TRANSDUCERS

Microphones

Several different types of microphone are available, the most common types used being condenser, dynamic and ribbon microphones.

Chloriona vasconica Rib.

Notus flavipennis (Zett.)

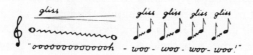

Euacanthus interruptus (L.)

Figure 11.1. Musical notation used by Ossiannilsson to describe the songs of 'small cicadas' or auchenorrynchous bugs. (From Ossiannilsson 1949).

Condenser or capacitance microphones consist of a thin membrane made from a conductor, such as aluminium, which is closely applied parallel to a rigid plate from which it is separated by a dielectric. This may be air or a plastic material. In some models the diaphragm is made from very thin aluminized plastic sheet, and the plastic backing forms the dielectric material. A DC voltage is applied via a fixed resistor between the diaphragm and the back plate which acts as a capacitor (Figure 11.2a). Vibration of the diaphragm causes a change in capacitance proportional to the displacement, which results in a fluctuating voltage through the circuit. These microphones have a very high output impedance, and a pre-amplifier stage, consisting of a field effect transistor (FET), is usually incorporated into the unit.

Condenser microphones have a reasonably high output and good dynamic response. They are small, robust and inexpensive. Their major potential disadvantage is that they require a power source to polarize the diaphragm. This is overcome in the electret microphone in which the dielectric is made from a material which can retain a charge for a long period. Unfortunately the charge is eventually lost and these microphones tend to change their characteristics with time. However, the development of small dry-cell batteries with long shelf lives has virtually negated this disadvantage, and condenser microphones are widely used in animal acoustics.

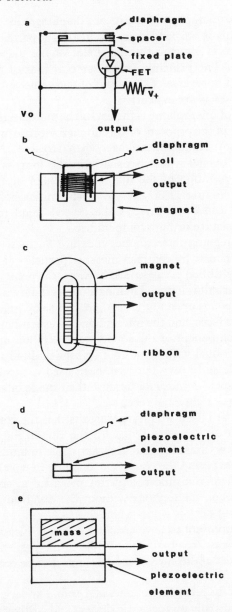

Figure 11.2. Diagrams to illustrate the principles of operation behind different types of transducer. a, Condenser or capacitance microphone; b, Dynamic microphone; c, Piezoelectric or crystal microphone; d, Ribbon microphone; e, Vibrometer for measuring substrate-borne vibrations. FET = field effect transistor.

The dynamic microphone consists of a diaphragm to which a light coil is attached. This is suspended in the field of a permanent magnet in which a voltage is induced by movement of the diaphragm and coil (Figure 11.2b). The principle is the same as that of the moving coil loudspeaker but in reverse. The output of these microphones is low and good quality models are expensive.

A third type of microphone is the crystal or piezoelectric microphone. This makes use of the property of certain piezoelectric materials, such as Rochelle salt, to produce a voltage when stressed. A diaphragm is coupled to the crystal which is usually in the form of a sandwich or bimorph, and vibration of the diaphragm results in a corresponding output voltage (Figure 11.2c). While crystal microphones have a high output – 0.5 V can be attained – the crystals tend to deteriorate if subjected to moisture or high temperature.

All of the foregoing microphones measure fluctuations in pressure. Ribbon microphones, by contrast, measure particle velocity. A narrow corrugated alloy ribbon is suspended between the poles of a permanent magnet and a current is induced in the ribbon as it moves (Figure 11.2d). Because the ribbon is open to the air there can be no pressure difference between the two faces, and the microphone is responding to the particle displacement component of the sound wave. Ribbon microphones are inherently directional with a cardioid or semicardioid response. They are rather fragile and heavy but have been used to record the songs of small insects such as *Drosophila* because their mode of action is similar to that of the insect antennal receptor.

A survey of the recent literature suggests that the condenser microphone is almost universally used, with the ribbon microphone being employed in a few cases. It is, however, possible to modify a condenser microphone by exposing the rear of the diaphragm which converts it to a pressure gradient transducer, and this provides an acceptable alternative to the ribbon microphone without the disadvantage of fragility (Bennet-Clark 1984).

Where it is important to have absolute intensity measurements, calibrated microphones are obtainable. While in the past these were often crystal types, most of the current models appear to be condenser microphones.

Hydrophones for recording underwater sound are available commercially. They are usually piezoelectric devices and, as they are designed to withstand high pressures and corrosion by sea water, tend to be bulky and expensive. However, for almost all insect recordings these features are unimportant and a simple hydrophone can be constructed using a condenser microphone insert encased in a thin sheath of material such as neoprene. Much of the difficulty in making underwater recordings, particularly in the laboratory, is sound reflections from the tank sides and the air–water interface. A general rule is to make use of as large a volume of water as practicable and to have the microphone as close to the sound source as possible.

Vibration transducers

An efficient vibration transducer must have a sufficiently low mass so that it does not damp out or distort the vibratory signals that are produced by small arthropods. The simplest method, and one that has been extensively employed, is to use a crystal or ceramic phonograph pick-up. Mono cartridges are now difficult to obtain but act in the vertical plane and are therefore ideal. They have a high output and there are no problems with impedance matching.

More sophisticated purpose-built vibrometers or acceleratometers contain a piezoelectric element which is acted upon by a seismic mass when the assembly is vibrated (Figure 11.2e). These instruments are reliable and have a wide frequency and dynamic range. While they suffer the same disadvantage as pickups, in that they load the vibrating structure, miniature accelerometers are obtainable which have a mass of under a gramme.

The most elaborate method of measuring vibration is to use laser vibrometry. A small spot of discrete light is focused on a reflective spot attached to the vibrating surface, which may be a part either of an animal or the substrate. The reflected light is doppler shifted by the vibrations and these can be recorded and analysed (Michelsen and Larsen 1978). This method is ideal as it does not interfere with the vibrating structure. However, the apparatus is expensive and not generally available.

RECORDING METHODS

The almost universal method of storing acoustic signals is with a tape recorder. Open-reel models have considerable advantages over cassette recorders, particularly for laboratory work. Most good quality open-reel recorders have several speeds, usually three or four, which can be between 2.4 and 38 mm s^{-1}, but a few possess even higher tape speeds. At the slowest speeds this can give over six hours of continuous recording at the expense of some fidelity and the loss of the higher frequencies. The ability to record at a high speed and play back at a lower one facilitates the location of particular signals for analysis. The one disadvantage of tape recorders designed for audio work is that they normally have an upper-frequency limit of around 20 kHz. This is because of the maximum tape speed available and the record head bias frequency which is usually between 50 and 100 kHz. Instrumentation tape recorders make use of much higher tape speeds and bias frequencies and will record frequencies well into the ultrasonic range. It is necessary therefore to use these expensive machines if one is interested in recording the songs of some bush-crickets and moths. 'Bat detectors' are available which trans-late ultrasonic signals into the sonic range. These retain the temporal characteristics of the signals and, as they can be tuned, give a rough idea concerning the carrier frequency of the signals (see Pye 1983 for more details).

Open-reel recorders tend to be bulky and many field workers now use cassette recorders. Some of these, such as the Sony WMD6C, have extremely good specifications and what they lack in versatility is com-pensated for by their portability.

A recent development has been the digital tape recorder which stores acoustic information in digitized form on video-type tape. These recorders should provide a better harmonic and dynamic range and an improved signal-to-noise ratio. It should also be possible to access the digitized information directly for analysis as is done routinely now for physiological recordings. At present these recorders are expensive but the price should fall dramatically if they are produced for playing music cassettes.

DISPLAY AND ANALYSIS

Acoustic signals can be analysed and displayed visually in the time or frequency domanes or, less satisfactorily, in both simultaneously. As the temporal parameters of arthropod signals are usually the more impor-tant, visualization provided by an oscilloscope screen is often adequate. Either still or moving cameras can be used to provide a hard copy for measurement and analysis. If the signals have a dominant carrier fre-quency this can be measured from oscilloscope pictures along with the temporal parameters.

An oscilloscope display cannot, however, provide much information on the spectral composition of signals which possess transients or strong harmonics. This type of signal is usually displayed by means of sonograms or audiospectrograms which are print-outs of frequency on the Y-axis against time on the X-axis. The density of the image provides a measure of the sound intensity. Earlier sonographs made use of a tape loop which was played repeatedly through a series of filters to provide the necessary information on frequency and intensity. This meant that only brief signals could be analysed. Modern instruments carry out an instantaneous fast Fourier transform (FFT) of the signal and much longer records can be analysed. Sonograms have been almost universally used to display bird and mammal vocalizations which are spectrally complex but they are less appropriate for many arthropod signals.

Intensity versus time displays are sometimes seen and are particularly useful in popular or semi-popular accounts, as they are much more easily 'read' than are sonograms, especially if accompanied by a verbal descrip-tion of the pitch of the sounds (Figure 11.3).

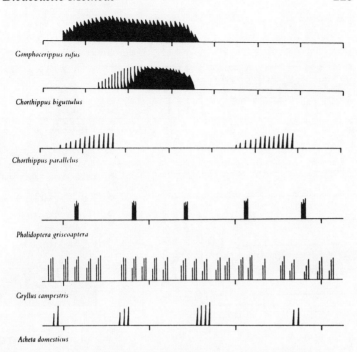

Figure 11.3. Stylized diagrams of amplitude versus time plots of the songs of various European Orthoptera. Time marker, 1 second. (From Bellmann 1988).

A more quantitative measure of spectral composition than that from a sonogram is provided by the spectrum analyzer. Any waveform can be approximated by the sum of a series of sine and cosine values and the method by which this is carried out is called Fourier analysis. A spectrum analyser carries out a FFT of a signal and displays the data graphically as a histogram of amplitude against frequency. This is known as the power spectrum of a signal. While spectrum analysers which are made for research and industrial use are expensive, it is now possible to purchase moderately priced digital oscilloscopes which possess the same facility and can be output to an X-Y plotter or to a printer, via a microcomputer.

To provide a direct comparison of some of these methods the court-ship song of *Drosophila melanogaster* and the calling song of *Gryllus bimaculatus* have been analysed on a Kay digital sonograph, model K7800. This provides a power spectrum, a sonogram and a time trace of the same signal. The results of these three types of analysis are illustrated in Figures 11.4 and 11.5.

The spread of microcomputers has opened up the possibility of quite complex analysis without the use of expensive specialist equipment. The author has an apparatus costing under £ 100, which, used in conjunction

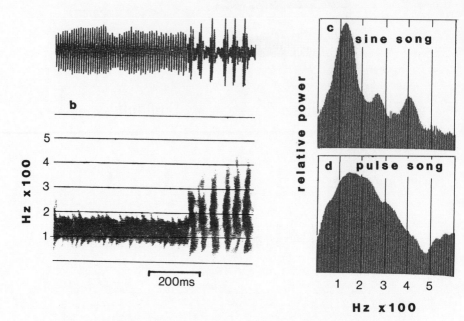

Figure 11.4. Simultaneous analysis of the song of *Drosophila melanogaster* to provide a comparison of different techniques. a, Oscillogram display; b, Sonogram; c, Power spectrum of sine song with much of the energy peaked at about 140 Hz and showing weak second and third harmonics; d, Power spectrum of pulse song. The energy is more widely spread than in sine song.

with his P.C., will digitize an acoustic signal, store it on disc and provide either a real-time picture or a power spectrum or an amplitude plot. These can then be dumped to a printer to provide a permanent record.

The problem for early bioacousticians was to develop methods of recording and analysing animal sounds. With the variety of methods now commonly available the difficulty is to choose a method which is appropriate to the particular problem under investigation.

PLAYBACK

Playback experiments have been extensively used to assess phonotactic responses and to provide simulated song. The simplest method is to use a tape loop of recordings of the actual song. Most open-reel recorders can be modified easily to play such loops. However, it is often desirable to be able to manipulate the songs and to alter individual parameters. This requires the use of synthesized songs. Appropriate synthesizers are not commercially available and have to be constructed. Circuit details have been published by Robinson and Ewing (1978) and Taylor (1979)

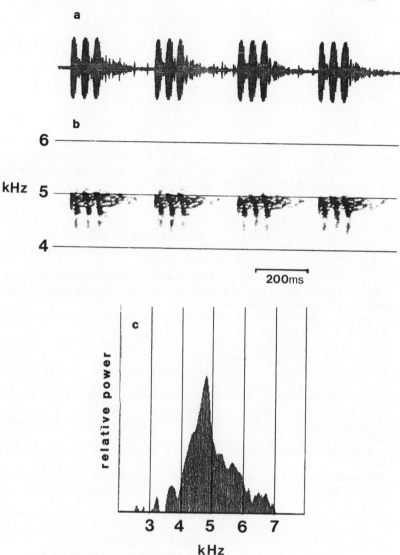

Figure 11.5. a, Oscillogram; b, Sonogram and c, Power spectrum of the calling song of *Gryllus bimaculatus*.

for song simulators designed for *Drosophila* and gryllid songs respectively.

In most of the recent experiments using artificial songs these are computer simulated and the parameters are controlled by computer programs. This reduces the necessity for quite complex electronic circuits which are replaced by computer software (Campbell and Forest 1987).

Glossary of bioacoustic terms

amplitude modulation: Periodic modulation of a carrier frequency so that the carrier frequency remains unchanged and the amplitude fluctuates.

bandpass filter: A circuit (electronic or neuronal) which allows the transmission of a specific range of frequencies.

c Propagation velocity of sound $[\mathrm{ms}^{-1}]$

In air at sea level $c \simeq 340\,\mathrm{ms}^{-1}$.

In fresh water $c \simeq 1430\,\mathrm{ms}^{-1}$. The exact value depends upon pressure, temperature and tonicity; c in sea water is usually taken to be $1500\,\mathrm{ms}^{-1}$.

carrier frequency: The underlying frequency of a signal before modification such as amplitude modulation.

chirp: A sound which may consist of one or more syllables or pulses and which is normally heard by the human ear as a unitary event. Usually applied to the songs of crickets and related insects.

decibel: dB; One tenth of a bel, a logarithmic unit used to compare two levels of power, i.e., $N\,dB = \log_{10}(p_2/p_1)$.

Note that N can be either negative or positive. An increase of 6 dB is the equivalent of doubling the power.

Doppler effect: Change in apparent frequency of a periodic signal such as a sound pressure wave or light due to movement of the emitter or receiver.

far field: Distance greater than one wavelength from a sound source. In this region pressure and particle velocity are in phase and behave as plane waves.

frequency: The number of cycles or oscillations per unit time in a periodic system.

frequency modulation: A perodic change in frequency of a carrier frequency.

fundamental: The frequency component of a note, usually the lowest, which provides the major proportion of the acoustic energy. Also called the first harmonic.

harmonic: A note whose frequency is an integral multiple of the fundamental frequency. This is only true of an idealized system. In real circumstances the harmonics diverge from the integral values to a varying extent.

high-pass filter: A circuit (electronic or neuronal) which only permits the transmission of high-frequency signals.

hertz: Hz; the SI unit of frequency. The number of cycles per second.

I Sound intensity, $[\mathrm{Wm}^{-2}]$

$I = pu = p^2/c\rho$, thus intensity varies with the square of the pressure $(I \propto p^2)$.

IL Intensity level: $10\log_{10} I/Ir\ dB$.

Ir Reference level for sound intensity. Usually taken to be $10^{-12}\,\mathrm{Wm}^{-2}$

J Joule: unit of energy $[\mathrm{Nm}^{-1}]$

low-pass filter: A circuit (electronic or neuronal) which only permits the transmission of low-frequency signals.

N Newton: unit of force $[\mathrm{kgs}^{-2}]$

near field: The area less than one wavelength from a sound source. In this area pressure and particle velocity are out of phase and sound measurements are unpredictable.

P Sound pressure [Pa or $\mathrm{Nm^{-2}}$]

Pa Pascal: SI unit of pressure, $1\,\mathrm{Pa} = 1\,\mathrm{Nm^{-2}}$

polar plot: A diagram drawn in polar coordinates to show the contours of equal energy surrounding either an emitter or receiver. Used, for example, to illustrate directionality of acoustic receptors.

power spectrum: Any complex waveform is composed of discrete sine waves in harmonic series. Fourier analysis is one method of analysing a sound to provide such a series, and a power spectrum is the result of such an analysis where the energy in each of the harmonic components is usually displayed graphically.

pr Sound pressure reference level: $20\,\mu\mathrm{Pa}$ $(2 \times 10^{-5}\,\mathrm{Nm^{-2}})$ (sometimes $2 \times 10^{-4}\,\mathrm{dynes\,cm^2}$)

pulse: A unitary sound normally produced by a single movement of the sound-producing apparatus. It may be monocyclic, as in the sound produced by the wing of some *Drosophila* species, or polycyclic as in the closing stridulatory wing stroke of crickets.

Q Quality factor: a measure of the degree of frequency tuning in an acoustic system. $Q = f^0/\Delta f$ (3 dB) where f_0 is the resonant frequency of the system.

resonant frequency: The frequency of an oscillating system at which the maximum amplitude of oscillation occurs in response to a driving force.

signal to noise ratio: The ratio of the energy contained in the parameter of a signal important for communication to that from irrelevant or extraneous sources.

SPL Sound pressure level measured in decibels relative to a reference level: $\mathrm{SPL} = 20\log_{10} p/\mathrm{pr}\,\mathrm{dB}$.

syllable: Synonymous with pulse.

tonotopic: In the nervous system; the map-like arrangement of acoustic inter-neurones such that their frequency responses and position are related.

transient: The sudden alteration of a signal due to the application or removal of a driving force.

trill: A sound in which the individual syllables run together to produce a more or less continuous note which may be amplitude-modulated.

u Vibration velocity [$\mathrm{ms^{-1}}$]: $u = p/c\rho$

ur Vibration velocity reference level: $0.5 \times 10^{-7}\,\mathrm{ms^{-1}}$

W Watt: unit of power [$\mathrm{Js^{-1}}$]

λ Wavelength: $\lambda \propto 1/f$; $\lambda = c/f$

ρ Acoustic density of medium [$\mathrm{kgm^{-3}}$]

ρ (air) $= 1.2\,\mathrm{kgm^{-3}}$

ρ (water) $= 1000\,\mathrm{kgm^{-3}}$

ω Angular frequency: $\omega = 2\pi$ radians

Arthropod groups from which acoustic behaviour has been reported

Phylum ARTHROPODA

Subphylum Crustacea
 Order Decapoda   <u>Pistol shrimps</u>
        <u>Fiddler crabs</u>

Subphylum Chelicerata
 Class Arachnida
  Order Scorpiones   <u>Scorpions</u>
  Order Arachnida   <u>Spiders</u>

Subphylum Uniramia
    Class Diplopoda   *Millipedes*
    Class Insecta
        Order   Odonata   *Dragonflies*
                       Orthoptera   *Crickets, grasshoppers**
                       Isoptera   *Termites*
                       Plecoptera   *Stoneflies*
                       Psocoptera   *Booklice*
                       Hemiptera   *Bugs*
                       Coleoptera   *Beetles*
                       Mecoptera   *Scorpion flies*
                       Tricoptera   *Caddis flies*
                       Lepidoptera   *Butterflies, moths*
                       Diptera   *Two-winged flies*
                       Siphonaptera   *Fleas*
                       Hymenoptera   *Bees, wasps*

The Hemiptera and Orthoptera are sometimes subdivided into several orders.

Order   Hemiptera
        Suborder   Homoptera   *Cicadas, planthoppers*
        Suborder   Heteroptera   *Shieldbugs, waterbugs*

Alternative classification
    Order Hemiptera ≡ Homoptera
    Order Heteroptera ≡ Heteroptera

Order Orthoptera
        Suborder Caelifera
        Family Acrididae   *Grasshoppers, locusts*
        Suborder Ensifera
        Family Tettigoniidae   *Bush-crickets or katydids,*
                                     long-horned grasshoppers

        Family Stenopelmatidae   *Wetas*
        Family Gryllidae   *Crickets*
        Family Gryllotalpidae   *Mole-crickets*
        Suborder Dictyoptera
        Family Mantidae   *Praying mantids*
        Family Blattidae   *Cockroaches*

Alternative classification
    Order Saltatoria ≡ Caelifera, Ensifera
    Order Mantoidea ≡ Mantidae
    Order Blattoidea ≡ Blattiae

A confusion exists with the common names of the Tettigoniidae. In North America these are variously called katydids, long-horned grasshoppers and cone-headed grasshoppers; in Europe they are all usually named bush-crickets. As they belong to the Ensifera, the European usage is probably preferable.

References

Adam, L.J. and Schwartzkopff, J. (1967) Getrennte nervose Reprasentation für verschiedene Tonbereiche im Protocerebrum von *Locusta migratoria*, *Z. vergl. Physiol.* 54, 6–255.

Agee, H.R. (1971) Ultrasound produced by wings of adults of *Heliothis zea*. *J. Insect Physiol.* 17, 1267–73.

Aicher, B., Markl, H., Masters, W.M. and Kirschenlohr, H.L. (1983) Vibration transmission through the walking legs of the fiddler crab, *Uca pugilator* (Brachyura, Ocypodidae) as measured by laser Doppler vibrometry, *J. Comp. Physiol.* 150, 483–91.

Aicher, B. and Tautz, J. 1984. 'Peripheral inhibition' of vibration-sensitive units in the leg of the fiddler crab *Uca pugilator*, *J. Comp. Physiol.* 154, 49–52.

Aidley, D.J. (1969) Sound production in a Brazilian cicada, *J. exp. Biol.* 51, 325–37.

Aiken, R.B. (1985) Sound production by aquatic insects, *Biol. Rev.* 60, 163–211.

Alcock, J. and Buchmann, S.L. (1985) The significance of post-inseminatory display by male *Centris pallida* (Hymenoptera: Anthrophoridae), *Zeit. für Tierpsychol.* 68, 231–43.

Alexander, A.J. (1958) On the stridulation of scorpions, *Behaviour* 12, 339–52.

Alexander, R.D. (1956) A comparative study of sound production in insects, with special reference to the singing Orthoptera and Cicadidae of the eastern United States. Ph.D. Thesis, Ohio State University, Columbus.

Alexander, R.D. (1960) Communicative mandible-snapping in Acrididae (Orthoptera), *Science* 132, 152–3.

Alexander, R.D. (1961) Aggressiveness, territoriality, and sexual behavior in field crickets (Orthoptera: Gryllidae), *Behaviour* 17, 130–223.

Alexander, R.D. (1962) Evolutionary change in cricket acoustical communication, *Evolution* 16, 443–67.

Alexander, R.D. (1975) Natural selection and specialized chorusing behavior in acoustical insects. In D. Pimental (ed.) *Insects, Science and Society*, pp. 35–77. New York: Academic Press.

Alexander, R.D. and Bigelow, R.S. (1960) Allochronic speciation in field crickets and a new species, *Acheta veletis*, *Evolution* 14, 334–46.

Alexander, R.D. and Moore, T.E. (1962) The evolutionary relationships of the 17-year and 13-year cicadas, and three new species (Homoptera, Cicadidae, *Magicicada*), *Misc. Pub. Mus. Zool. University of Michigan* 121, 1–59.

Alexander, R.D., Moore, T.E. and Woodruff, R.E. (1963) The evolutionary differentiation of stridulatory signals in beetles (Insecta: Coleoptera), *Anim. Behav.* 11, 111–5.

Anderson, M. (1978) Spirally arranged muscles associated with tracheoles in tsetse fly flight muscles: their possible involvement in sound production, *Experientia* 34, 587–9.

Asahina, V.S., (1939) Tonerzeugung bei *Epiophlebia*-Larven (Odonata, Anisozygoptera), *Zool. Anz.* 126, 323–5.

Assem, van den J. and Putters, F.A. (1980) Pattern of sound produced by courting chalcidoid males and its biological significance, *Entomol Exp. Appl.* 27, 293–302.

Atkins, G., Ligman, S., Burghardt, F. and Stout, J.F. (1984) Changes in phonotaxis by the female cricket *Acheta domesticus* L. after killing identified acoustic interneurons, *J. Comp. Physiol.* 154, 795–804.

Autrum, H. (1941) Uber Gehor- und Erschutterungssinn bei Locustiden. *Z. vergl. Physiol.* 28, 580–637.

Bailey, W.J. (1970) The mechanics of stridulation in bush crickets (Tettigonioidea, Orthoptera). I. The tegminal generator, *J. exp. Biol.* 52, 495–505.

Bailey, W.J. (1978) Resonant wing systems in the Australian whistling moth *Hecatesia* (Agarasidae, Lepidoptera), *Nature* 272, 444–6.

Bailey, W.J. and Broughton, W.B. (1970) The mechanics of stridulation in bush crickets (Tettigoinoidea, Orthoptera). II. Conditions for resonance in the tegminal generator, *J. exp. Biol.* 52, 507–17.

Bailey, W.J. and Thomson, P. (1977) Acoustic orientation in the cricket *Teleogryllus oceanicus* (Le Guillou), *J. exp. Biol.* 67, 61–75.

Barth, F.G. (1982) Spiders and vibratory signals. In P.N. Witt and J.S. Rovner (eds) *Spider Communication*, pp. 67–122. Princeton, N.J.: Princeton University Press.

Barth, F.G. and Geethabali (1982) Spider vibration receptors: threshold curves of individual slits in the metatarsal lyriform organ, *J. Comp. Physiol.* 148, 175–85.

Bauer, M. and von Helversen, O. (1987) Separate localization of sound recognizing and sound producing neural mechanisms in a grasshopper, *J. Comp. Physiol.* 161, 95–101.

Bell, P.D. (1979) Acoustic attraction of herons by crickets, *J.N.Y. Entomol. Soc.* 87, 126–7.

Bellman, H. (1988) *A Field Guide to the Grasshoppers and Crickets of Britain and Northern Europe*, London: Collins.

Belton, P. (1962) Responses to sound in pyralid moths, *Nature, Lond.* 196, 1188–9.

Belwood, J.J. and Morris, G.K. (1987) Bat predation and its influence on calling behavior in neotropical katydids, *Science* 238, 64–7.

Bennet-Clark, H.C. (1970) The mechanism and efficiency of sound production in mole crickets, *J. Exp. Biol.* 52, 619–52.

Bennet-Clark, H.C. (1971) Acoustics of insect song. *Nature* 234, 255–9.

Bennet-Clark, H.C. (1975) Sound production in insects. *Sci. Prog. Oxf.* 62, 263–83.

Bennet-Clark, H.C. (1984) A particle velocity microphone for the song of small insects and other near-field measurements, *J. exp. Biol.* 108, 459.

Bennet-Clark, H.C. and Ewing, A.W. (1967) Stimuli provided by courtship of male *Drosophila melanogaster*, *Nature* 215, 669–71.

Bennet-Clark, H.C. and Ewing, A.W. (1968) The wing mechanism involved in the courtship of *Drosophila*, *J. exp. Biol.* 49, 117–28.

Bennet-Clark, H.C. and Ewing, A.W. (1969) Pulse interval as a critical parameter in the courtship song of *Drosophila melanogaster*, *Anim. Behav.* 17, 755–9.

Bennet-Clark, H.C., Leroy, Y. and Tsacas, L. (1980) Species and sex-specific songs and courtship behaviour in the genus *Zaprionus* (Diptera-Drosophilidae), *Anim. Behav.* 28, 230–55.

Bentley, D.R. (1969) Intracellular activity in cricket neurons during generation of song patterns, *Z. vergl. Physiol.* 62, 267–83.

Bentley, D.R. (1977) Control of cricket song patterns by descending interneurons. *J. comp. Physiol.* 116, 19–38.

Bentley, D.R. and Hoy, R.R. (1970) Postembryonic development of adult motor patterns in crickets: a neural analysis, *Science* 170, 1409–11.

Bentley, D.R. and Hoy, R.R. (1974) The neurobiology of cricket song, *Scientific American* 231, 34–44.

Bentley, D.R. and Kutsch, W. (1966) The neuromuscular mechanisms of stridulation in crickets (Orthoptera: Gryllidae), *J. exp. Biol.* 45, 151–64.

Blest, A.D. (1964) Protective display and sound production in some new world arctiid and ctenuchid moths, *Zoologica* 49, 161–81.

Blest, A.D., Collett, T.S. and Pye, J.D. (1963) The generation of ultrasonic signals by a New World arctiid moth, *Proc. Roy. Soc. B.* 158, 196–207.

Blum, M.S. and Blum, N.A. (1979) Eds. *Sexual Selection and Reproductive Competition in Insects*, New York: Academic Press.

Boake, C.R.B. and Capranica, R.R. (1982) Aggressive signal in 'courtship' chirps of a gregarious cricket, *Science* 218, 580–2.

Boyan, G.S. (1981) Two-tone suppression of an identified auditory neurone in the brain of the cricket *Gryllus bimaculatus* (De Geer), *J. Comp. Physiol.* 144, 117–25.

Boyan, G.S. and Altman, J.S. (1985) The suboesophageal ganglion: a 'missing link' in the auditory pathway of the locust. *J. Comp. Physiol.* 156, 413–28.

Boyan, G.S. and Williams, J.L.D. (1982) Auditory neurones in the brain of the cricket *Gryllus bimaculatus* (De Geer): ascending interneurones, *J. Insect. Physiol.* 28, 493–501.

Boyd, P., Kuhne, R., Silver, S. and Lewis, B. (1984) Two-tone suppression and song coding by ascending neurones in the cricket *Gryllus campestris* L., *J. Comp. Physiol.* 154, 423–30.

Boyd, P. and Lewis, B. (1983) Peripheral auditory directionality in the cricket (*Gryllus campestris* L., *Teleogryllus oceanicus* Le Guillou). *J. Comp. Physiol.* 152, 523–32.

Bradbury, J.W. and Andersson, M.B. (1987) *Sexual Selection: Testing the Alternatives*. Chichester: Wiley.

Brownell, P.H. (1977) Compressional and surface waves in sand used by desert scorpions to locate prey, *Science* 197, 479–82.

Bruinsma, O., Kruijt, J.P. and Dusseldorp, W. van. (1981) Delay of emergence of honey bee queens in response to tooting sounds, *Proc. Kon. Med. Akad. van Wetensch. C.* 84, 381–7.

Buchler, E.R., Wright, T.B. and Brown, E.D. (1981) On the functions of stridulation by the passalid beetle *Odontotaenius disjunctus* (Coleoptera: Passalidae), *Anim. Behav.* 29, 483–6.

Burk, T. (1982) Evolutionary significance of predation on sexually signalling males, *Fla. Entomol.* 65, 90–104.

Burk, T. and Webb, J.C. (1983) Effect of male size on calling propensity: song parameters and mating success in caribbean fruit flies, *Anastrepha suspensa* (Diptera: Tephritidae), *Ann. Entomol. Soc. Am.* 76, 678–82.

Burnet, B., Connolly, K. and Dennis, L. (1971) The function and processing of auditory information in the courtship of *Drosophila melanogaster*, *Anim. Behav.* 19, 409–5.

Burrows, M. (1973) The role of delayed excitation in the co-ordination of some metathoracic flight motoneurons of a locust, *J. Comp. Physiol.* 83, 135–64.

Busnel, R.G. and Dumortier, B. (1960) Vérification par des méthodes d'analyse acoustique des hypothèses sur l'origine du cri du Sphinx *Acherontia atropos* (L), *Bull. Soc. Entomol. Fr.* 64, 44–58.

Busnel, R.G., Dumortier, B. and Busnel, M.-C. (1956) Recherches sur le comportement acoustique des Ephippigères (Orthopteres, Tettigoniidae), *Bull. Biol. Fr. Belg.* 15, 219–86.

Butlin, R.K. (1987) Speciation by reinforcement. *Trends Ecol. Evol.* 2, 8–13.

Butlin, R.K. and Hewitt, G.M. (1986) The response of female grasshoppers to male song, *Anim. Behav.* 34, 1896–9.

Butlin, R.K., and Hewitt, G.M. (1987) The structure of grasshopper song in relation to mating success, *Behaviour* 104, 152–61.

Butlin, R.K. and Hewitt, G.M. (1988) Genetics of behavioural and morphological differences between parapatric subspecies of *Chorthippus parallelus* (Orthoptera: Acrididae), *Biol. J. Linn. Soc.* 33, 233–48.

Butlin, R.K., Hewitt, G.M. and Webb, S.F. (1985) Sexual selection for intermediate optimum in *Chorthippus brunneus* (Orthoptera: Acrididae), *Anim. Behav.* 33, 421–9.

Cade, W. (1975) Acoustically orienting parasitoids: Fly phonotaxis to cricket song, *Science* 190, 1312–3.

Cade, W.H. (1981a) Field cricket spacing, and the phonotaxis of crickets and parasitoid flies to clumped and isolated cricket songs, *Z. Tierpsychol.* 55, 365–75.

Cade, W.H. (1981b) Alternative male strategies: genetic differences in crickets, *Science* 212, 563–4.

Cade, W.H. and Wyatt, D.R. (1984) Factors affecting calling behaviour in field crickets, *Teleogryllus* and *Gryllus* (age, weight, density and parasites), *Behaviour* 88, 61–75.

Campbell, D.J. and Clarke, D.J. (1971) Nearest neighbour test of significance for non-randomness in the spatial distribution of singing crickets (*Teleogryllus commodus* (Walker)), *Anim. Behav.* 19, 750–6.

Campbell, D.J. and Forest, J. (1987) A cricket-song simulator using the 67805 single-chip microcomputer, *Behav. Res. Instrum. Comput.* 19, 26–9.

Carde, R.T. and Baker, T.C. (1984) Sexual communication with pheromones. In W.J. Bell and R.T. Carde (eds) *Chemical Ecology of Insects*. pp. 355–63. London: Chapman and Hall.

Carlson, A.D., Copeland, J., Raderman, R. and Bulloch, A.G.M. (1976) Role of interflash intervals in firefly courtship (*Photinus macdermotti*), *Anim. Behav.* 24, 786–92.

Caruso, D. and Costa, G. (1976) L'apparato stridulatore e l'emissione di suoni in *Armidillo officinalis* Dumeril (Crustacea, Isopoda, Oniscoidea), *Animalia* 3, 17–27.

Chjvala, M. Diskpcil, J., Mook, J.H. and Pokorny, B. (1974) The genus *Lipara* Meigen (Diptera, Chloropidae), systematics, morphology, behaviour, and ecology, *Tijdsch. voor Entomol.* 117, 1–25.

Claridge, M.F. (1974) Stridulation and defensive behaviour in the ground beetle, *Cychrus caraboides* (L), *J. Ent. (A)* 49, 7–16.

Claridge, M.F. (1985) Acoustic signals in the Homoptera: behavior, taxonomy and evolution, *Ann. Rev. Entomol.* 30, 297–317.

Claridge, M.F., den Hollander, J. and Morgan, J.C. (1984) Specificity of acoustic signals and mate choice in the brown planthopper *Nilaparvata lugens*, *Entomol. exp. appl.* 35, 221–6.

Claridge, M.F., den Hollander, J. and Morgan, J.C. (1988) Variation in hostplant relations and courtship signals of weed-associated populations of the brown planthopper, *Nilaparvata lugens* (Stal), from Australia and Asia: a test of the recognition species concept, *Biol. J. Linn. Soc.* 35, 79–93.

Claridge, M.F. and Reynolds, W.J. (1973) Male courtship songs and sibling species in the *Oncopsis flavicollis* species group (Hemiptera: Cicadellidae), *J. Entomol. B.* 42, 29–39.

Cobb, M., Connolly, K. and Burnet, B. (1985) Courtship behaviour in the *melanogaster* species sub-group of *Drosophila*, *Behaviour* 95, 203–31.

Coggshall, J.C. (1978) Neurons associated with the dorsal longitudinal flight muscles of *Drosophila melanogaster*, *J. Comp. Neur.* 177, 707–20.

Cokl, A. (1983) Functional properties of vibroreceptors in the legs of *Nezara viridula* (L.) (Heteroptera, Pentatomidae), *J. Comp. Physiol.* 150, 261–9.

Cokl, A. (1985) Problems of sound communication in a land bug species *Nezara viridula* L. (Heteroptera, Pentatomidae) In K. Klamring and N. Elsner (eds) *Acoustic and Vibrational Communication in Insects*, pp. 163–8, Berlin and Hamburg: Paul Parey.

Cokl, A., Kalmring, K. and Wittig, H. (1977) The responses of auditory ventral cord neurons of *Locusta migratoria* to vibration stimuli, *J. Comp. Physiol.* 120, 161–72.

Cowling, D.E. (1980) The genetics of *Drosophila melanogaster* courtship song – Diallel analysis, *Heredity* 45, 401–3.

Cowling, D.E. and Burnet, B. (1981) Courtship songs and genetic control of their acoustic characteristics in sibling species of the *Drosophila melanogaster* subgroup, *Anim. Behav.* 29, 924–35.

Coyne, J.A. and Orr, H.A. (1989) Patterns of speciation in *Drosophila, Evolution*, 43, 362–381.

Crankshaw, O.S. (1979) Female choice in relation to calling and courtship song in *Acheta domesticus, Anim. Behav.* 27, 1274–5.

Dadour, I.R. and Bailey, W.J. (1985) Male agonistic behaviour of the bushcricket *Mygalopsis marki* Bailey in response to conspecific song (Orthoptera: Tettigoniidae), *Z. Tierpsychol.* 70, 320–30.

Darwin, C. (1871) *The Descent of Man and Selection in Relation to Sex*, New York: Modern Library.

Dethier, V.G. (1963) *The Physiology of Insect Senses*, London: Methuen.

Dish, V.M. (1961) A preliminary revision of the families and subfamilies of acridoidea (Orthoptera, Insecta), *Bull. Brit. Mus. (Nat. Hist.) Ent.* 10, 351–419.

Doherty, J.A. (1985a) Phonotaxis in the cricket, *Gryllus bimaculatus* DeGeer: comparisons of choice and no-choice paradigms, *J. Comp. Physiol.* 157, 279–89.

Doherty, J.A. (1985b) Trade-off phenomena in calling song recognition and phonotaxis in the cricket, *Gryllus bimaculatus* (Orthoptera, Gryllidae), *J. Comp. Physiol.* 156, 787–801.

Doherty, J.A. (1985c) Temperature coupling and 'trade-off' phenomena in the acoustic communication of the cricket, *Gryllus bimaculatus* De Geer (Gryllidae), *J. exp. Biol.* 114, 17–35.

Doherty, J.A. and Gerhardt, H.C. (1983) Hybrid tree frogs: vocalizations of males and selective phonotaxis of females, *Science* 220, 1078–80.

Downes, J.A. (1969) The swarming and mating flight of Diptera, *Ann. Rev. Entomol.* 14, 271–98.

Dumortier, B. (1963) Etude experimentale de la valeur interspécifique du signal acoustique chez les éphippigères et rapport avec les problèmes d'isolement et de maintien de l'espèce, *Ann. Epiphyties* 14, 5–23.

Dumortier, B. (1971) Stridulation without stridulatory apparatus: an original method in Pamphaginae (Insecta, Orthoptera) *Forme et functio* 4, 265–9.

Dunning, D.C. (1968) Warning sounds of moths, *Zeit. für Tierpsychol.* 25, 129–38.

Dunning, D.C., Byers, J.A. and Zanger, C.D. (1979) Courtship in two species of periodical cicadas, *Magicicada septendecim* and *Magicicada cassini, Anim. Behav.* 27, 1073–90.

Dunning, D.C. and Roeder, K.D. (1965) Moth sounds and the insect catching behaviour of bats, *Science* 147, 173–4.

Dybas, H.S. and Davis, D.D. (1962) A population census of seventeen-year periodic cicadas (Homoptera: Cicadidae: *Magicicada*), *Ecology* 43, 432–44.

Eibl, E. (1974) Verlauf des Hornerven und Projekt der im 'Hornerven' verlaufenden Fesern im Prothorakalganglion von Grillen mit Hilfe der CoS-Methode. Cologne: Staatsarbeit.

Eibl, E. (1978) Morphology of the sense organs in the proximal parts of the tibiae of *Gryllus campestris* L. and *Gryllus bimaculatus* de Geer (Insecta, Ensifera), *Zoomorphologie* 89, 185–205.

Eisner, T., Aneshansley, D., Eisner, M., Rutowski, B., Chong, B. and Meinwald, J. (1974) Chemical defense and sound production in Australian tenebrionid beetles (*Adelium* spp.), *Psyche* 81, 189–208.

Elepfandt, A. (1980) Morphology and output coupling of wing muscle motoneurons in the field cricket (Gryllidae, Orthoptera), *Zool. Jb. (Physiol.)* 84, 26–45.

Elephandt, A. and Popov, A.V. (1979) Auditory interneurones in the mesothoracic ganglion of crickets. *J. Insect Physiol.* 25, 429–41.

Elliot, C.J.H. (1983) Wing hair plates in crickets: physiological characteristics and connections with stridulatory motor neurones, *J. exp. Biol.*
107, 21–47.

Elliott, C.J.H. and Koch, U.T. (1985) The clockwork cricket, *Naturwissenschaften* 72, 150–2.

Elsner, N. (1970) Die Registrierung der Stridulationsbewegungen bei der
Feldheuschrecke *Chorthippus mollis* mit Hilfe Hallgeneratoren, *Z. Vergl.
Physiol.* 68, 417–28.

Elsner, N. (1973) The central nervous control of courtship behaviour in the
grasshopper *Gomphocerippus rufus* L. (Orthoptera: Acrididae). In J.
Salanki, (Ed.) *Neurobiology of Invertebrates*, pp. 261–87. Budapest:
Akademiai Kiado.

Elsner, N. (1984) Neuroethology of sound production in gomphocerine
grasshoppers. I. Song patterns and stridulatory movements, *J. Comp.
Physiol.* 88, 72–102.

Elsner, N. (1983) A neuroethological approach to the phylogeny of leg stri-
dulation in gomphocerine grasshoppers. In F. Huber and H. Markl (eds)
Neuroethology and Behavioural Physiology, pp. 54–68, Berlin: Springer-
Verlag.

Elsner, N. and Huber, F. (1969) Die Organisation des Werbegesangs der
Heuschrecke *Gomphocerippus rufus* L. in Abhangigkeit von zentralen und
peripheren Bedingungen, *Z. vergl. Physiol.* 65, 389–423.

Elsner, N. and Popov, A.J. (1978) Neuroethology of acoustic communication,
Advances in Insect Physiology 13, 229–355.

Esch, H. (1963) Auswirkung der Futterplatzqualität auf die Schallerzeugung
im Werbetanz der Honigbiene, *Verh. Dtsch. Zool. Ges.* 1962, 302–9.

Esch, H. (1967) Die Bedeutung der Lauterzeugung für die Verstandgung der
stachellosen Bienen, *Z. vergl. Physiol.* 56, 199–220.

Esch, H., Huber, F. and Wohlers, D.W. (1980) Primary auditory neurons in
crickets: physiology and primary projections, *J. Comp. Physiol.* 137, 27–
35.

Ewing, A.W. (1961) Body size and courtship behaviour in *Drosophila melanogaster*, *Anim. Behav.* 9, 93–9.

Ewing, A.W. (1964) The influence of wing area on the courtship behaviour
of *Drosophila melanogaster*, *Anim. Behav.* 12, 316–20.

Ewing, A.W. (1969) The genetic basis of sound production in *Drosophila
pseudoobscura* and *D. persimilus*, *Anim. Behav.* 17, 555–60.

Ewing, A.W. (1970) The evolution of courtship songs in *Drosophila*, *Rev.
Comport. Anim.* 4, 3–8.

Ewing, A.W. (1977) The neuromuscular basis of courtship song in *Drosophila*: the role of the indirect flight muscles, *J. Comp. Physiol.* 119,
249–65.

Ewing, A.W. (1978) The antenna of *Drosophila* as a 'love song' receptor,
Physiol. Entomol. 3, 33–6.

Ewing, A.W. (1979a) The neuromuscular basis of courtship song in *Drosophila*: the role of the direct and axillary wing muscles, *J. Comp. Physiol.*
130, 87–93.

Ewing, A.W. (1979b) The role of feedback during singing and flight in
Drosophila melanogaster, *Physiol. Entomol* 4, 329–37.

Ewing, A.W. (1979c) Complex courtship songs in the *Drosophila funebris*
species group: escape from an evolutionary bottleneck, *Anim. Behav.* 27,
343–9.

Ewing, A.W. (1983) Functional aspects of *Drosophila* courtship, *Biol. Rev.*
58, 275–92.

Ewing, A.W. and Bennet-Clark, H.C. (1968) The courtship songs of
Drosophila, *Behaviour* 31, 288–301.

Ewing, A.W. and Hoyle, G. (1965) Neuronal mechanisms underlying control of sound production in a cricket: *Acheta domesticus*, *J. exp. Biol.* 43, 139–53.

Ewing, A.W. and Miyan, J.A. (1986) Sexual selection, sexual isolation and the evolution of song on the *Drosophila repleta* group of species, *Anim. Behav.* 34, 421–9.

Field, L.H. (1978) The stridulatory apparatus of New Zealand wetas in the genus *Hemideina* (Insecta: Orthoptera: Stenopelmatidae), *J. Roy. Soc. New Zealand*, 8, 359–75.

Field, L.H., Evans, A. and MacMillan, D.L. (1987) Sound production and stridulatory structures in hermit crabs of the genus *Trizopagurus*, *J. Mar. Biol., U.K.* 67, 89–110.

Field, L.H., Hill, K.G. and Ball, E.E. (1980) Physiological and biophysical properties of the auditory system of the New Zealand weta *Hemideina crassidens* (Blanchard, 1851) (Ensifera: Stenopelmatidae), *J. Comp. Physiol.* 141, 31–7.

Finke, C. and Prager, J. (1980) Pulse-train synchronous pair-stridulation by male *Sigara striata* (Heteroptera, Corixidae), *Experientia* 36, 1172–3.

Fisher, R.M. and Weary, D.M. (1988) Buzzing bees: Communication between bumble bee social parasites (Hymenoptera: Apidae) and their hosts, *Bioacoustics* 1, 3–12.

Fletcher, N.H. (1978) Acoustical response of hair receptors in insects, *J. Comp. Physiol.* 127, 185–9.

Fletcher, N.H. and Hill, K.G. (1978) Acoustics of sound production and of hearing in the bladder cicada *Cystosoma saudersii* (Westwood), *J. exp. Biol.* 72, 43–55.

Forrest, T.G. (1983) Calling song and mate choice in mole crickets. In D.T. Gwynne and G.K. Morris (eds) *Orthopteran Mating Systems* pp. 185–204, Boulder, Colorado: Westview Press.

Fraser, J. and Nelson, M.C. (1982) Frequency modulated courtship song in a cockroach, *Anim. Behav.* 30, 637–8.

Freitag, R. and Lee, S.K. (1972) Sound producing structures in *Cicindela tranquebarica* (Coleoptera, Cicindellidae) including a list of tiger beetles and ground beetles with flight wing files, *Can. Ent.* 104, 851–7.

Fuchs, S. (1976) The response to vibrations of the substrate and reactions to the specific drumming in colonies of carpenter ants (*Camponotus*, Formicidae, Hymenoptera), *Behav. Ecol. Sociobiol.* 1, 155–84.

Fulton, B.B. (1928) A demonstration of the location of auditory organs in certain Orthoptera, *Ann. Entomol. Soc. Am.* 21, 445–8.

Fulton, B.B. (1933) Inheritance of songs in hybrids of two subspecies of *Nemobius fasciatus* (Orthoptera). *Ann. Entomol. Soc. Am.* 26, 368–76,.

Gogala, M. (1984) Vibration producing structures and songs of terrestrial Heteroptera as a systematic character, *Biol. Vest.* 32, 19–36.

Gogala, M. (1985) Vibrational communication in insects. In K. Kalmrig and N. Elsner (eds) *Acoustic and Vibrational Communication in Insects*, Berlin: Paul Parey.

Greenfield, M.D. (1983) Unsynchronized chorusing in the conehead katydid *Neoconocephalus affinis*, *Anim. Behav.* 31, 102–12.

Greenfield, M.D. and Shelly, T.E. (1985) Alternative mating strategies in a desert grasshopper: evidence for density-dependence, *Anim. Behav.* 33, 1192–210.

Guthrie, D.M. (1966) Sound production and reception in a cockroach, *J. exp. Biol.* 45, 321–8.

Gwynne, D.T. (1982) Mate selection by female katydids (Orthoptera: Tettigoniidae, *Conocephalus nigropleurum*), *Anim. Behav.* 30, 734–8.

Gwynne, D.T. and Bailey, W.J. (1988) Mating system, mate choice and ultrasonic calling in a zaprochiline katydid (Orthoptera: Tettigoniidae), *Behaviour* 105, 202–23.

Gwynne, D.T. and Morris, G.K. (1986) Heterospecific recognition and behavioural isolation in acoustic Orthoptera (Insecta), *Evol. Theory* 8, 33–8.

Hagen, H.-O. von (1984) Visual and acoustic display in *Uca mordax* and *U. burgesi*, sibling species of neotropical fiddler crabs. II. Vibrational signals, *Behaviour* 91, 204–28.

Hall, J. (1985a) Neuroanatomical and neurophysiological aspects of vibrational processing in the central nervous system of semi-terrestrial crabs. I. Vibration sensitive interneurons in the fiddler crab, *Uca minax, J. Copmp. Physiol.* 157, 91–104.

Hall, J. (1985b) Neuroanatomical and neurophysiological aspects of vibrational processing in the central nervous system of semi-terrestrial crabs. II. Comparative anatomical and physiological aspects of stimulus processing, *J. Comp. Physiol.* 157, 105–13.

Hamilton, W.D. (1971) Geometry of the selfish herd, *J. theor. Biol.* 31, 295–311.

Hannemann, H.J. (1956) Über ptero-tarsale Stridulation und einige andere Arten Lauterzeugung bei Lepidopteren, *Deutsch. Entomol. Zeit.* 3, 14–27.

Harrison, J.B. (1969) Acoustic behaviour of a wolf spider, *Lycosa gulosa, Anim. Behav.* 17, 14–6.

Hartley, J.C. and Robinson, D.J. (1976) Acoustic behaviour of both sexes of the speckled bush cricket *Leptophyes punctatissima, Physiol. Entomol* 1, 21–5.

Haskell, P.T. (1957) Stridulation and its analysis in certain Geocorisae (Hemiptera Heteroptera), *Proc. Zool. Soc. Lond.* 129, 351–8.

Haskell, P.T. and Belton, P. (1956) Electrical responses of certain lepidopterous tympanal organs, *Nature, Lond.* 177, 139–40.

Hassenstein, B. (1959) Optokinetisch Wirksamkeit bewegter periodischer Muster, *Z. Naturf.* 14, 659–74.

Hawkins, A.D. and Myrberg, A.A. (1983) Hearing and sound communication under water. In B. Lewis (ed.) *Bioacoustics, a comparative approach*, pp. 347–406, London: Academic Press.

Hazlett, B.A. and Wynn, H.E. (1962) Sound production and associated behavior of Bermuda crustaceans *(Palinurus, Gonodactylus, Alpheus* and *Synalpheus), Crustaceana* 4, 25–38.

Heady, S.E., Nault, L.R. Shambaugh, G.F. and Fairchild, L. (1986) Acoustic and mating behavior of *Dalbulus* leafhoppers (Homoptera: Cicadellidae), *Ann. Entomol. Soc. Am.* 79, 727–36.

Hedrick, A.V. (1986) Female preferences for male calling bout duration in a field cricket, *Behav. Ecol. Sociobiol.* 19, 73–7.

Hedrick, A.V. (1988) Female choice and the heritability of attractive male traits: an empirical study, *Am. Nat.* 132, 267–76.

Hedwig, B. (1986) On the role in stridulation of plurisegmental interneurons of the acridid grasshopper *Omocestus viridulus* L. II. Anatomy and physiology of ascending and T-shaped interneurons, *J. Comp. Physiol.* 158, 429–44.

Hedwig, B. and Elsner, N. (1980) A neuroethological analysis of sound production in the acridid grasshopper *Omocestus viridulus, Adv. Physiol. Sci.* 23, 495–514.

Hedwig, B. and Elsner, N. (1985) Sound production and sound detection in a stridulating acridid grasshopper *(Omocestus viridulus).* In K. Kalmring and N. Elsner (eds) *Acoustic and Vibrational Communication in Insects*, Berlin and Hamburg: Paul Parey.

Heide, G. (1983) Neural mechanisms of flight control in Deptera. In W. Nachtigall (ed.) *Biona-report 2*, pp. 35–52. Stuttgart: Gustav Fischer.

Heinrich, B. (1979) *Bumblebee Economics*, Cambridge, Mass.: Harvard University Press.

Heinzel, H.-G. and Dambach, M. (1987) Travelling air vortex rings as potential communication signals in a cricket, *J. Comp. Physiol.* 160, 79–88.

Heller, K.-G. and Helversen, D. von (1986) Acoustic communication in phaneropterid bushcrickets: species:specific delay of female stridulatory response and matching male sensory time window, *Behav. Ecol. Sociobiol.* 18, 189–98.

Helversen, O. von and Elsner, N. (1977) The stridulatory movements of acridid grasshoppers recorded with an opto-electronic device. *J. Comp. Physiol.* 122, 53–64.

Helversen, D. von and Helversen, O. von (1983) Species recognition and acoustic localization in acridid grasshoppers: a behavioral approach. In F. Huber and H. Markl (eds) *Neuroethology and Behavioral Physiology*, pp. 97–107, Berlin: Springer-Verlag.

Helversen, O. von and Helversen, D. von, (1987) Innate receiver mechanisms in the acoustic communication of orthopteran insects. In D.M. Guthrie (ed.) *Aims and Methods in Neuroethology*, Manchester: University Press.

Hemming, A.J. and Jansson, A. (1980) Acoustic behavior of *Geotrupes stercorarius* L. (Coleoptera, Geotrupidae). *3rd Euro. Meet. Bioacoustic Insects.*

Hennig, R.M. (1988) Ascending auditory interneurons in the cricket *Teleogryllus commodus* (Walker): comparative physiology and direct connections with afferents, *J. Comp. Physiol.* 163, 135–44.

Henry, C.S. (1979) The courtship of *Chrysopa downsei* Banks (Neuroptera: Chrysopidae): its evolutionary significance, *Psyche* 86, 291–7.

Henry, C.S. (1980) The importance of low-frequency, substrate-borne sounds in lacewing communication (Neuroptera: Chrysopidae), *Ann. ent. Soc. Amer.* 73, 617–21.

Henry, C.S. (1983) Acoustic recognition of sibling species within the holarctic lacewing *Chrysoperla carnea* (Neuroptera: Chrysopidae), *Syst. Entomol.* 8, 293–301.

Hertwick, M. 1931. Anatomie und Variabilität des Nervensystems und der sinnesorgane von *Drosophila melanogaster*, *Z. wiss. Zool.* 139, 559–63.

Hill, K.G. and Boyan, G.S. (1977) Sensitivity to frequency and direction of sound in the auditory system of crickets (Gryllidae), *J. Comp. Physiol.* 121, 79–97.

Hill, K.G., Loftus-Hills, J.J. and Gartside, D.F. (1972) Premating isolation between the Australian field crickets *Teleogryllus commodus* and *T. oceanicus* (Orthoptera: Gryllidae), *Aust. J. Zool.* 20, 153–63.

Hill, K.G. and Oldfield, B.P. (1981) Auditory function in Tettigoniidae (Orthoptera: Ensifera), *J. Comp. Physiol.* 142, 169–80.

Hinton, H.E. (1948) Sound production in lepidopterous pupae, *Entomologist* 81, 254–69.

Hinton, H.E. (1955) Protective devices of endopterygote pupae, *Trans. Soc. Brit. Entomol.* 12, 49–92.

Hograefe, T. (1984) Substratum-stridulation in the colonial sawfly larvae of *Hemichroa crocea* (Hymenoptera: Tenthridinidae), *Zool. Anz.* 213, 234–41.

Hoikkala, A., Lakovaara, S. and Romppainen, E. (1982) Mating behavior and male courtship sounds in the *Drosophila virilis* group. In S. Lakovarra (ed.) *Advances in Genetics, Development and Evolution of Drosophila*, pp. 407–21, New York: Plenum.

Hoikkala, A. and Lumme, J. (1984) Genetic control of the difference in male courtship sound between *Drosophila virilis* and *D. lummei*, *Behav. Genet.* 14, 257–68.

Hoikkala, A. and Lumme, J. (1987) The genetic basis of evolution of the male courtship sounds in the *Drosophila virilis* group, *Evolution* 41, 827–45.

Horridge, F.A. (1961) Pitch discrimination in locusts, *Proc. R. Soc. Lond. B.* 155, 218–31.

Howard, D.J. and Furth, D.G. (1986) Review of the *Allonemobius fasciatus* (Orthoptera: Gryllidae) complex with descriptions of two new species separated by electrophoresis, songs and morphometrics, *Ann. Entomol. Soc. Am.* 79, 472–81.

Howe, M.A. and Lehane, M.J. (1986) Post-feed buzzing in the tse-tse, *Glossina morsitans morsitans*, is an endothermic mechanism, *Physiol. Entomol.* 11, 279–86.

Howse, P.E. (1964) The significance of the sound produced by the termite *Zootermopsis angusticollis* (Hagen), *Anim. Behav.* 12, 284–300.

Hoy, R.R., Hahn, J. and Paul, R.C. (1977) Hybrid cricket auditory behavior: evidence for genetic coupling in animal communication, *Science* 195, 82–4.

Hoy, R.R., Hoikkala, A. and Kaneshiro, K. (1988) Hawaiian courtship song: evolutionary innovation in communication signals of *Drosophila*, *Science* 240, 217–9.

Hoy, R.R. and Paul, R.C. (1973) Genetic control of song specificity in crickets, *Science* 180, 82–3.

Huber, F. (1955) Sitz und Bedeutung nervoser Zentren für Instinkthandlungen beim Mannchen von *Gryllus campestris* L. *Z. Tierpsychol.* 12, 12–48.

Huber, F. (1960) Untersuchungen über die Funktion des Zentralnervensystems und insbesondere des Gehirns bei der Fortbewegung und der Lauterzeugung der Grillen, *Z. vergl. Physiol.* 44, 60–132.

Huber, F. (1978) The insect nervous system and insect behaviour, *Anim. Behav.* 26, 969–81.

Huber, F. (1983) Neural correlates of orthopteran and cicada phonotaxis. In F. Huber, and H. Markl (eds) *Neuroethology and Behavioral Physiology*. Berlin: Springer-Verlag.

Huber, F. and Thorson, J. (1985) Cricket auditory communication, *Scientific American* 253, 60–8.

Huber, F., Wohlers, D.W. and Moore, T.E. (1980) Auditory nerve and interneurone responses to natural sounds in several species of cicadas, *Physiol. Entomol* 3, 25–45.

Huntingford, F. and Turner, A. (1987) *Animal Conflict*, London: Chapman and Hall.

Hutchings, M. and Lewis, B. (1984) The role of two-tone suppression in song coding by ventral cord neurones in the cricket *Teleogryllus oceanicus* (Le Guillou), *J. Comp. Physiol.* 154, 103–12.

Ikeda, H. and Maruo, O. (1982) Directional selection for pulse repetition rate of the courtship sound and correlated responses occurring in several characters in *Drosophila mercatorum*, *Jpn. J. Genet.* 57, 241–58.

Ikeda, H., Takabatake, I. and Sawada, N., (1980) Variation in courtship sound among three geographical strains of *Drosophila mercatorum*, *Behav. Gen.* 10, 361–75.

Jansson, A. (1973) Stridulation and its significance in the genus *Cenocorixa* (Hemiptera, Corixidae), *Behaviour* 46, 1–36.

Jansson, A. (1975) Mounting signals and their function in three *Sigara (Subsigara)* species (Hemiptera, Corixidae), *Abs. XIV Int. Ethol. Conf.*

Jansson, A. (1976) Audiospectrographic analysis of stridulatory signals of some North American Corixidae (Hemiptera). *Ann. Zool. Fennici* 13, 48–62.

Jansson, A. (1979) Reproductive isolation and experimental hybridization between *Arctocorisa carinata* and *A. germari* (Heteroptera, Corixidae), *Ann. Zool. Fennici* 16, 89–104.

Jansson, A. and Vuoristo, T. (1979) Significance of stridulation in larval Hydropsychidae (Trichoptera), *Behaviour* 71, 167–86.

Johnson, M.W., Alton Everest, F. and Young R.W. (1947) The role of snapping shrimp (*Crangon* and *Synalpheus*) in the production of underwater noise in the sea, *Biol. Bull.* 93, 122–38.

Jones, M.D.R. (1966) The acoustic behaviour of the bush cricket *Pholidoptera griseoaptera*. I. Alternation, synchronysm and rivalry between males, *J. exp. Biol.* 45, 15–30.

Josephson, R.K. and Halverson, R.C. (1971) High frequency muscles used in sound production by a Katydid. I. Organization of the motor system, *Biol. Bull.* 141, 411–433.

Kalmring, K. (1975) The afferent auditory pathway in the ventral cord of *Locusta migratoria* (Acrididae). II. Response of the auditory ventral cord neurons to natural sounds, *J. Comp. Physiol.* 104, 143–59.

Kalmring, K. and Kuhne, R. (1980) The coding of airborne sound and vibration signals in bimodal ventral-cord neurons of the grasshopper *Tettigonia cantans, J. Comp. Physiol.* 139, 267–75.

Kalmring, K., Kuhne, R. and Lewis, B. (1983) The acoustic behaviour of the bushcricket *Tettigonia contans* III. Coprocessing of auditory and vibratory information in the central nervous system, *Behav. Proc.* 8, 213–28.

Kalmring, K. Kuhne, R. and Moysich, R. (1978) The coding of sound signals in the ventral-cord auditory system of the migratory locust, *Locusta migratoria, J. Comp. Physiol.* 128, 213–226.

Kalmring, K., Lewis, B. and Eichendorf, A. (1978) The physiological characteristics of the primary sensory neurons of the complex tibial organ of *Decticus verrucivorus* L. (Orthoptera, Tettigonioidae). *J. Comp. Physiol.* 127, 109–21.

Kammer, A.E. (1981) Physiological mechanisms of thermoregulation. In B. Heinrich (ed.) *Insect Thermoregulation*, pp. 115–58, New York: Wiley.

Kamper, G. (1985) Processing of species-specific low-frequency song components by interneurons in crickets. In Kalmring, K. and Elsner, N. (eds) *Acoustic and Vibrational Communication in Insects*, Berlin and Hamburg: Paul Parey.

Kamper, G. and Dambach, M. (1979) Communication by infrasound in a non-stridulating cricket. *Naturwissenschaften* 66, 530.

Kamper, G. and Dambach, M. (1981) Responses of the cercus-to-giant interneuron system in crickets to species-specific song, *J. Comp. Physiol.* 141, 311–7.

Karban, R. (1982) Increased reproductive success at high densities and predation satiation for periodic cicadas, *Ecology* 63, 321–8.

Kavanagh, M.W. (1987) The efficiency of sound production in two cricket species, *Gryllotalpa australis* and *Teleogryllus commodus* (Orthoptera: Grylloidea), *J. exp. Biol.* 130, 107–19.

Kawanishi, M. and Watanabe, T.K. (1980) Genetic variation of courtship song of *Drosophila melanogaster* and *D. simulans, Japan J. Genet.* 55, 235–40.

Kawanishi, M. and Watanabe, T.K. (1981) Genes affecting courtship song and mating preference in *Drosophila melanogaster, Drosophila simulans* and their hybrids, *Evolution* 35, 1128–33.

Keiser, I., Kobayashi, R.M., Chambers, D.L. and Schneider, E.L. (1973) Relation of sexual dimorphism in the wings, potential stridulation and illumination to mating of oriental fruit flies, melon flies and Mediterranean fruit flies in Hawaii, *Ann. Entomol. Soc. Am.* 66, 937–41.

Keuper, A., Weidemann, S., Kalmring, S. and Kaminiski, D. (1988) Sound production and sound emission in seven species of European tettigoniids. Part I. The different parameters of the song; their relation to the morphology of the bushcricket, *Bioacoustics* 1, 26–31.

Kevan, D.C.McE. (1954) Méthodes inhabituelle de production de son chez les orthoptères. In R.G. Busnel (ed.) *L'acoustique des Orthoptères*, pp. 103–41, Paris: INRA.

Kinsler, L.E., Frey, A.R., Coppens, A.B. and Sanders, J.V. (1982) *Fundamentals of Acoustics*, New York: Wiley.

Kleindienst, H.U., Wohlers, D.W. and Larsen, O.N. (1983) Tympanal membrane motion is necessary for hearing in crickets, *J. Comp. Physiol.* 151, 397–400.

Koenig, J.H. and Ikeda, K. (1983) Characterization of the intracellularly recorded response of identified flight motor neurons in *Drosophila, J. Comp. Physiol.* 150, 295–303.

Konichi, M. (1965) The role of auditory feedback in the control of vocalization in the white-crowned sparrow, *Z. Tierpsychol.* 22, 770–83.

Krafft, B. (1978) The recording of vibrational signals performed by spiders during courtship, *Symp. Zool. Soc. Lond.* 42, 59–67.

Kuhne, R. (1982) Neurophysiology of the vibration sense in locusts and bushcrickets: response characteristics of single receptor units, *J. Insect Physiol.* 28, 155–63.

Kuhne, R., Silver, S. and Lewis, B. (1984) Processing of vibratory acoustic signals by ventral cord neurones in the cricket *Gryllus campestris, J. Insect Physiol.* 30, 575–85.

Kutsch, W. (1969) Neuromuskulare Aktivität bei verschiedenen Verhaltensweisen von drei Grillenarten, *Z. vergl. Physiol.* 63, 335–78.

Kutsch, W. and Otto, D. (1972) Evidence for song production independent of head ganglia in *Gryllus campestris. J. Comp. Physiol.* 81, 115–9.

Larsen, O.N. and Michelsen, A. (1978) Biophysics of the ensiferan ear III. The cricket ear as a four-input system, *J. Comp. Physiol.* 123, 217–27.

Latimer, W. and Schatral, A. (1986) Information cues used in male competition by *Tettigonia cantans* (Orthoptera: Tettigoniidae), *Anim. Behav.* 34, 162–8.

Latimer, W. and Sippel, M. (1987) Acoustic cues for female choice and male competition in *Tettigonia cantans, Anim. Behav.* 35, 887–900.

Lee, R.C.P., (1986) The role of the male song in sex recognition in *Zaprionus tuberculatus* (Diptera, Drosophilidae), *Anim. Behav.* 34, 641–48.

Leroy, Y. (1964) Analyse de la transmission des divers paramèters des signaux acoustiques chez les hybrides interspécifiques de grillons (Orthoptères, Ensifères). *Proc. XII Int. Cong. Ent.* 239.

Leroy, Y. (1965) Transmission du paramètre fréquence dans le signal acoustique des hybrides F1 et PxF1, de deux grillons: *Teleogryllus commodus* Walter et *T. oceanicus* Le Guillou (Orthoptères, Ensifères), *C.R. Acad. Sc. Paris* 259, 892–5.

Lewis, D.B. (1974) The physiology of the tettigoniid ear. III. The response characteristics of the intact ear and some biophysical considerations, *J. exp. Biol.* 60, 853–9.

Lloyd, J.E. (1977) Bioluminescence and communication in insects. In T.A. Sebeok (ed.) *How Animals Communicate*, pp. 164–83, Indiana Univ. Press.

Loher, W. (1957) Untersuchungen über den Aufbau und die Enstehung der Gesange einiger Feldheuschreckenarten und dein Einfluss von Lautzeichen auf das akuctische Verhalten, *Z. vergl. Physiol.* 39, 313–56.

Loher, W. and Chandrashekaran, M.K. (1972) Communicative behavior of the grasshopper *Syrbula fuscovittata* (Thomas) (Gomphocerinae) with particular consideration of the male courtship, *Z. Tierpsychol.* 31, 78–97.

Loher, W. and Huber, F. (1966) Nervous and endocrine control of sexual behavior in a grasshopper (*Gomphocerus rufus* L.), *Symp Soc. exp. Biol.* 20, 381–400.

McDonald, J. and Crossley, S. (1982) Behavioural analysis of lines selected for wing vibration in *Drosophila melanogaster, Anim. Behav.* 30, 802–10.

McVean, A. (1986) The song of the New Zealand weta, *Hemideina thoracica* (Orthoptera: Stenopelmatidae), *J. Zool. Lond. (A)* 208, 171–90.

Mangold, J.R. (1978) Attraction of *Euphasiopteryx ochracea, Corethrella* sp. and gryllids to broadcast songs of the southern mole cricket, *Fla. Entomol* 61, 57–61.

Manning, A. (1967a) Antennae and sexual receptivity in *Drosophila melanogaster* females, *Science* 158, 136–7.

Manning, A. (1967b) The control of sexual receptivity in female *Drosophila, Anim. Behav.* 15, 239–50.

Markl, H. (1973) The evolution of stridulatory communication in ants. *Proc. 7th. Congr. IUSSI*, London, 258–65.

Markl, H. (1983) Vibrational communication. In F. Huber and H. Markl (eds) *Neuroethology and Behavioral Physiology* Berlin: Springer-Verlag.

Markl, H. and Holldobler, B. (1978) Recruitment and foot-retrieving behavior in *Novomessor* (Formicidae, Hymenoptera). II. Vibrational signals. *Behv. Ecol. Sociobiol.* 4, 183–216.

Markl, H., Holldobler, B. and Holldobler, T. (1977) Mating behavior and sound production in harvester ants (Pogonomyrmex, Formicidae), *Insectes Sociaux* 24, 191–212.

Masters, W.M. (1979) Insect disturbance stridulation: Its defensive role, *Behav. Ecol. Sociobiol.* 5, 187–200.

Masters, W.M. and Markl, H. (1981) Vibrational signal transmission in spider orb webs, *Science* 213, 363–5.

Masters, W.M., Tautz, J., Fletcher, N.H. and Markl, H. (1983) Body vibration and sound production in an insect (*Atta sexdens*) without specialized radiating structures, *J. Comp. Physiol.* 150, 239–49.

Mather, K. and Jinks, J.L. (1982) *Biometrical Genetics* 3rd ed., London: Chapman and Hall.

Mayr, E. (1972) Sexual selection and natural selection. In B. Campbell (ed.) *Sexual Selection and the Descent of Man, 1871–1971*, pp. 87–104. Chicago: Aldine.

Meixner, A.J. and Shaw, K.C. (1979) Spacing and movement of singing *Neoconocephalus nebrascensis* males (Tettigoniidae: Copophorinae), *Ann. Entomol. Soc. Am.* 72, 602–6.

Meixner, A.J. and Shaw, K.C. (1986) Acoustic and associated behavior of the coneheaded katydid, *Neoconocephalus nebrascensis* (Orthoptera: Tettigoniidae), *Ann. Entomol. Soc. Am.* 79, 554–65.

Michel, K. (1974) Das Tympanalorgan von *Gryllus bimaculatus* deGeer Saltatoria, Gryllidae), *Z. Morphol. Tiere* 77, 285–315.

Michelsen, A. (1971a) The physiology of the locust ear I. Frequency sensitivity of single cells in the isolated ear, *Z. vergl. Physiol* 71, 49–62.

Michelsen, A. (1971b) The physiology of the locust ear II. Frequency discrimination based upon resonances in the tympanum, *Z. vergl. Physiol.* 71, 63–101.

Michelsen, A. (1971c) The physiology of the locust ear III. Accoustical properties of the intact ear, *Z. vergl. Physiol.* 71, 102–28.

Michelsen, A. (1978) Sound reception in different environments. In. M.A. Ali (ed.) *Sensory Ecology*, New York: Plenum.

Michelsen, A. (1983) Biophysical basis of sound communication. In B. Lewis (ed.) *Bioacoustics, a Comparative Approach*, London: Academic Press.

Michelsen, A. (1985) Environmental aspects of sound communication in insects. In K. Kalmring and N. Elsner (eds) *Acoustic and Vibrational Communication in Insects*, Berlin: Paul Parey.

Michelsen, A., Fink, F., Gogala, M. and Traue, D. (1982) Plants as transmissions channels for insect vibrational songs, *Behv. Ecol. Sociobiol.* 11, 269–81.

Michelsen, A., Kirchner, W.H., Andersen, B.B. and Lindauer, M. (1986) The tooting and quacking vibration signals of honeybee queens: a quantitative analysis, *J. Comp. Physiol.* 158, 605–11.

Michelsen, A., Kirchner, W.H. and Lindauer, M. (1986) Sound and vibrational signals in the dance language of the honey bee, *Apis mellifera*, *Behav. Ecol. Sociobiol.* 18, 207–12.

Michelsen, A. and Larsen, O.N. (1978) Biophysics of the ensiferal ear I. Tympanal vibrations in bushcrickets (Tettigoniidae) studied with laser vibrometry, *J. Comp. Physiol.* 123, 193–203.

Michelsen, A. and Larsen, O.N. (1983) Strategies for acoustic communication in complex environments. In F. Huber and H. Markl (eds) *Neuroethology and Behavioral Physiology* Berlin: Springer-Verlag.

Michelsen, A. and Nocke, H. (1977) Biophysical aspects of sound communication in insects, *Adv. Insect. Physiol.* 10, 247–96.

Miller, D.D., Goldstein, R.B. and Patty, R.A. (1975) Semispecies of *Drosophila athabasca* distinguishable by male courtship sounds, *Evolution* 29, 531–44.

Miller, L.A. (1970) Structure of the green lacewing tympanal organ (*Chrysopa carnea*, Neuroptera), *J. Morph.* 131, 359–82.

Miller, L.A. (1971) Physiological responses of green lacewings (*Chrysopa carnea*, Neuroptera) to ultrasound, *J. insect Physiol.* 17, 491–506.

Miller, L.A. (1977) Directional hearing in the locust *Schistocerca gregaria* Forskal (Acrididae, Orthoptera), *J. Comp. Physiol.* 119, 85–98.

Miller, L.A. (1983) How insects detect and avoid bats. In F. Huber and H. Markl (eds) *Neuroethology and Behavioral Physiology*, Berlin: Springer-Verlag.

Miyan, J.A. and Ewing, A.W. (1984) A wing synchronous receptor for the dipteran flight motor, *J. Insect. Physiol.* 30, 567–74.

Miyan, J.A. and Ewing, A.W. (1985a) How Diptera move their wings: a re-examination of the wing base articulation and muscle systems concerned with flight, *Phil Trans. R. Soc. Lond. B* 311, 271–302.

Miyan, J.A. and Ewing, A.W. (1985b) Is the 'click' mechanism of dipteran flight an artefact of CCl_4 anaesthesia? *J. exp. Biol.* 116, 313–22.

Mohl, B. and Miller, L.A. (1976) Ultrasonic clicks produced by the peacock butterfly: a possible bat-repellent mechanism, *J. exp. Biol.* 64, 639–44.

Moiseff, A. and Hoy, R. (1983) Sensitivity to ultrasound in an identified auditory interneuron in the cricket: a possible neural link to phonotactic behavior, *J. Comp. Physiol.* 152, 155–67.

Moiseff, A., Polack, G. and Hoy, R. (1978) Steering responses of flying crickets to sound and ultrasound: mate attraction and predator avoidance, *Proc. Nat. Acad. Sci.* 75, 4052–56.

Mook, J.H. and Bruggemann, C.G. (1968) Acoustical communication by *Lipara lucens* (Diptera, Chloropidae), *Ent. Exp. and Appl.* 11, 397–402.

Morris, G.K. (1980) Calling, display and mating behavior of *Copiphora rhinoceros* Pictet (Orthoptera: Tettigoniidae), *Anim. Behav.* 28, 42–51.

Morris, G.K. and Fullard, J.H. (1983) Random noise and congeneric discrimination in *Conocephalus* (Orthoptera: Tettigoniidae). In D.T. Gwynne and G.K. Morris (eds) *Orthopteran Mating Systems: Sexual Competition in a Diverse Group of Insects.* pp. 73–96. Boulder: Westview Press.

Morris, G.K., Kerr, G.E. and Fullard, J.H. (1978) Phonotactic preferences of female meadow katydids (Orthoptera: Tettigoniidae: *Conocephalus nigropleurum*), *Can. J. Zool.* 56, 1479–87.

Morris, G.K. and Pipher, R.E. (1967) Tegminal amplifiers and spectrum consistencies in *Conocephalus nigropleurum* (Bruner), Tettigoniidae, *J. Insect Physiol.* 13, 1075–85.

Moss, D. (1971) Sinnesorgane im Bereich des Flügels der Feldgrille (*G. campestris*) und ihre Bedeutung für die Einstellung der Flügellage, *J. vergl. Physiol.* 73, 53–83.

Moulton, J.M. (1957) Sound production in the spiny lobster, *Panulirus argus, Biol. Bull* 113, 286–95.

Murphey, R.K., Palka, J. and Hustert, R. (1977) The cercus-to-giant interneuron system of crickets II. Response characteristrics of two giant interneurons, *J. Comp. Physiol.* 119, 285–300.

Murphey, R.K. and Zaretsky, M.D. (1972) Orientation to calling song by female crickets *Scapsipedus marginatus* (Gryllidae), *J. exp. Biol.* 56, 335–352.

Nairns, P.M. and Capranica, R.R. (1976) Sexual differences in the auditory system of the tree frog *Eleutherodactylus coqui, Science* 192, 378–80.

Nelson, M.C. (1979) Sound production in the cockroach, *Gromphadorhina portentosa*: The sound-producing apparatus, *J. Comp. Physiol.* 132, 27–38.

Nocke, H. (1971) Biophysik der Schallerzeugung durch die Vorderflügel der Grillen, *Z. vergl. Physiol.* 74, 272–314.

Nocke, H. (1972) Physiological aspects of sound communication in crickets (*Gryllus campestris*), *J. Comp. Physiol.* 80, 141–62.

Nocke, H. (1975) Physical and physiological properties of the tettigoniid 'Grasshopper' ear, *J. Comp. Physiol.* 100, 25–57.

Oldfield, B.P. (1982) Tonotopic organisation of auditory receptors in Tettigoniidae (Orthoptera: Ensifera), *J. Comp. Physiol.* 147, 461–69.

Oldfield, B.P. (1983) Central projections of primary auditory fibres in Tettigoniidae (Orthoptera: Ensifera), *J. Comp. Physiol.* 151, 389–95.

Oldfield, B.P. and Hill, K.G. (1983) The physiology of ascending auditory interneurons in the tettigoniid *Caedicia simplex* (Orthoptera: Ensifera): response properties and a model of integration in the afferent auditory pathway, *J. Comp. Physiol.* 152, 495–508.

Oldfield, B.P., Kleindienst, H.U. and Huber, F. (1986) Physiology and tonotopic organization of auditory receptors in the cricket *Gryllus bimaculatus* De Geer, *J. Comp. Physiol.* 159, 457–64.

Ossiannilsson, F. (1949) Insect drummers, *Opuscula Entomologica* Supp. 10, 145.

Otte, D. and Loftus-Hills, J. (1979) Chorusing in *Syrbula* (Orthoptera: Acrididae). Cooperation, interference competition or concealment?, *Entomol. News* 90, 159–65.

Otto, D. (1967) Untersuchungen zur nervosen Kontrolle des Grillengesangges. Verh. D. Zool. Ges. Heidelberg 1967, *Zool. Anz. Suppl.* 341, 585–92.

Otto, D. (1971) Untersuchungen zür zentralnervosen Kontrolle der Lauterzeugung von Grillen, *Z. vergl. Physiol.* 74, 227–71.

Otto, D. and Weber, T. (1982) Interneurons descending from the cricket cephalic ganglia that discharge in the pattern of two motor rhythms. *J. Comp. Physiol.* 148, 209–19.

Palka, J., Levine, R. and Schubiger, M. (1977) The cercus-to-giant interneuron system of crickets I. Some attributes of the sensory cells, *J. Comp. Physiol.* 119, 267–83.

Palka, J. and Olberg, R. (1977) The cercus-to-giant interneuron system of crickets. III. Receptive field organization, *J. Comp. Physiol.* 119, 301–17.

Papi, F. (1969) Light emission, sex attraction and male flash dialogues in a firefly, *Luciola lustanica* (Charp), *Monot. Zool. Ital.* 3, 135–84.

Partridge, L., Ewing, A.W. and Chandler, A. (1987) Male size and mating success in *Drosophila melanogaster*: the roles of male and female behavior, *Anim. Behav.* 35, 555–62.

Paterson, H.E.H. (1985) The recognition concept of species, *Transvaal Mus. Monogr.* 4, 21–9.

Paul, R.C. (1976) Species specificity in the phonotaxis of female ground crickets (Orthoptera: Gryllidae: Nemobiniinae), *Ann. Entomol. Soc. Am.* 69, 1007–10.

Pearman, J.V. (1928) On sound production in the Psocoptera and on a presumed stridulatory organ, *Entomol. Mon. Mag.* 14, 179–86.

Perdeck, A.C. (1957) The isolating value of specific song patterns in two sibling species of grasshoppers (*Chorthippus brunnerus* Thunb. and *C. biguttulus* L.), *Behaviour* 12, 1–75.

Phillips, L.H. and Konishi, M. (1973) Control of aggression by singing in crickets, *Nature* 241, 64–5.

Pierce, G.W. (1948) *The Songs of Insects*, Harvard Press, U.S.A.

Pollack, G.S. (1982) Sexual differences in cricket calling song recognition, *J. Comp. Physiol.* 146, 217–22.

Pollack, G.S. (1986) Discrimination of calling song models by the cricket, *Teleogryllus oceanicus*: the influence of sound direction on neural encoding of the stimulus temporal pattern and on phonotactic behavior, *J. Comp. Physiol.* 158, 549–61.

Pollack, G.S. and Hoy, R.R. (1981) Phonotaxis to individual rhythmic components of a complex cricket *Teleogryllus oceanicus* calling song, *J. Comp. Physiol.* 144, 367–74.

Pollack, G.S., Huber, F. and Weber, T. (1984) Frequency and temporal pattern-dependent phonotaxis of crickets (*Teleogryllus oceanicus*) during tethered flight and compensated walking, *J. Comp. Physiol.* 154, 13–26.

Popov, A.V. (1981) Sound production and hearing in the cicada *Cicadetta sinuatipennis* Osh. (Homoptera, Cicadidae), *J. Comp. Physiol* 142, 271–80.

Popov, A.V. and Markovich, A.M. (1982) Auditory interneurones in the prothoracic ganglion of the cricket, *Gryllus bimaculatus*, *J. Comp. Physiol.* 146, 351–9.

Popov, A.V. and Shuvalov, V.F. (1977) Phonotactic behavior of crickets, *J. Comp. Physiol.* 119, 111–26.

Popov, A.V., Shuvalov, V.F., Svetlogorskaya, I.D. and Markovich, A.M. (1974) Acoustic behavior and auditory system in insects. In J. Schwartzkopff (ed.) *Mechanoreception. Abh. Rhein.-Westf. Akad. Wiss.* 53, 281–306.

Prager, J. and Larsen, O.N. (1981) Asymmetrical hearing in the water bug *Corixa punctata* III. observed with laster vibrometry, *Naturwissenschaften* 68, 579–80.

Prager, J. and Streng, R. (1982) The resonance properties of the physical gill of *Corixa punctata* and their significance in sound reception, *J. Comp. Physiol.* 148, 323–35.

Prell, H. (1920) Die Stimme des Totenkopfes (*Acherontia atrops* L.), *Zoolog. Jahrb. Systm. Geog. Biol. Tiere.* 42, 235–72.

Prestwich, K.N. and Walker, T.J. (1981) Energetics of singing in crickets: Effect of temperature in three trilling species (Orthoptera: Gryllidae), *J. Comp. Physiol.* 143, 199–212.

Pringle, J.W.S. (1954) A physiological analysis of cicada song, *J. exp. Biol.* 32, 525–60.

Prozesky-Schulze, L., Prozesky, O.P.M., Anderson, F. and van der Merwe, G.J.J. (1975) Use of a self-made sound baffle by a tree cricket, *Nature* 255, 142–3.

Pye, J.D. (1983). Techniques for studying ultrasound. In B. Lewis (ed.) *Bioacoustics* pp. 39–65. London: Academic Press.

Rainlender, Y., Shuvalov, V.F., Popov, A.V. and Kalmring, L. (1981) Patterns of movement of female crickets *Gryllus bimaculatus* toward the source of the attracting signal and dependence of precision of orientation on the spectrum of the signal, *J. evol. Biochem. Physiol.* 17, 20–6.

Réaumur, R.A.F. de (1734) *Mémoires pour Servir à l'Histoire des Insectes*, 1. Paris.

Réaumur, R.A.F. de (1740) *Mémoires pour Servir à l'Histoire des Insectes*, 5. Paris.

Rehbein, H.G., Kalmring, K. and Romer, H. (1974) Structure and function of acoustic neurons in the thoracic ventral nerve cord of *Locusta migratoria* (Acrididae), *J. Comp. Physiol.* 95, 263–80.

Riede, K. (1983) Influence of the courtship song of the acridid grasshopper *Gomphocerus rufus* L. on the female, *Behav. Ecol. Sociobiol.* 14, 21–7.

Riede, K., (1987) A comparative study of mating behavior in some neotropical grasshoppers (Acridoidea), *Ethology* 76, 265–96.

Rindorf, H.J. (1981) Acoustic emission source location in theory and practice, *Bruel and Kjaer Tech. Rev.* 2, 3–32.

Robertson, R.M. and Pearson, K.G. (1985) Neural circuits in the flight system of the locust, *J. Neurophysiol.* 53, 110–28.

Robinson, D.J. and Ewing, A.W. (1978) An animal sound simulator, *Behav. Res. Met. Instrument.* 10, 848–51.

Robinson, D.J., Rheinlaender, J. and Hartley, J.C. (1986) Temporal parameters of male–female sound communication in *Leptophyes punctatissima*, *Physiol. Entomol.* 11, 317–23.

Roeder, K.D. (1966) Auditory system of noctuid moths. *Science* 154, 1515–8.

Roeder, K.D. and Treat, A.E. (1957) Ultrasonic reception by the tympanic organ of noctuid moths, *J. exp. Biol.* 134, 127–58.

Roeder, K.D. and Wallman, J. (1966) Directional sensitivity of the ears of noctuid moths, *J. exp. Biol.* 44, 17–31.

Romer, H. and Bailey, W.J. (1986) Insect hearing in the field. II. Male spacing behavior and correlated acoustic cues in the bushcricket *Mygalopsis marki*, *J. Comp. Physiol.* 159, 627–38.

Romer, H. and Dronse, R. (1982) Synaptic mechanisms of monaural and binaural processing in the locust, *J. Insect Physiol.* 28, 365–70.

Romer, H. and Marquart, V. (1984) Morphology and physiology of auditory interneurons in the metathoracic ganglia of the locust, *J. Comp. Physiol.* 155, 249–62.

Romer, H., Rheinlaender, J. and Dronse, R. (1981) Intracellular studies on auditory processing in the metathoracic ganglion of the locust, *J. Comp. Physiol.* 144, 305–12.

Romer, H. and Seikowski, U. (1985) Responses to model songs of auditory neurons in the thoracic ganglia and brain of the locust, *J. Comp. Physiol.* 156, 845–60.

Ronacher, B., Helversen, D. von and Helversen, O. von (1986) Routes and stations in the processing of auditory directional information in the CNS of a grasshopper as revealed by surgical experiments, *J. Comp. Physiol.* 158, 363–42.

Roth, L.M. (1948) A study of mosquito behavior, *Amer. Midland Nat.* 40, 265–352.

Rothschild, M. (1985) British aposematic lepidoptera. In J. Heath and A.M. Emmet (eds) *The Moths and Butterflies of Great Britain and Ireland, vol 2.* pp. 9–62. Colchester: Harley Books.

Rovner, J.S. (1975) Sound production by nearctic wolf spiders: A substratum-coupled stridulatory mechanism, *Science* 190, 1309–10.

Rovner, J.S. and Barth, F.G. (1981) Vibratory communication through living plants by a tropical wandering spider, *Science* 214, 464–6.

Rudinsky, J.A. and Ryker, L.C. (1976) Sound production in Scotylidae: rivalry and premating stridulation of male douglas-fir beetle, *J. Insect Physiol.* 22, 997–1003.

Rupprecht, R. (1976) Struktur und Funktion der Bauchblase und der Hammers von Plecopteren, *Zool. Jb. Anat.* 95, 9–80.

Ryan, M.J., Tuttle, M.D. and Taft, L.K. (1981) The costs and benefits of frog chorusing behavior, *Behav. Ecol. Sociobiol.* 8, 273–87.

Ryker, L.C. (1976) Acoustic behavior of *Tropisternus ellipticus, T. columbianus* and *T. lateralis limbalis* in western Oregon (Coleoptera: Hydrophilidae), *Coleopt. Bull.* 30, 147–56.

Saini, R.K. (1985) Sound production associated with sexual behaviour of the tsetse, *Glossina morsitans morsitans, Insect Sci. Applic.* 6, 637–44.

Sakaluk, S.K. and Belwood, J.J. (1984) Gecko phonotaxis to cricket calling song: A case of satellite predation, *Anim. Behav.* 32, 659–62.

Sales, F. and Pye, D. (1974) *Ultrasonic Communication by Animals* London: Chapman and Hall.

Salmon, M. and Horch, K.W. (1972) Acoustic signalling and detection by semiterrestrial crabs of the family Ocypodidae. In H.E. Winn and B.L. Olla (eds) *Behavior of Marine Animals*, vol. 1, pp. 61–96. New York: Plenum Press.

Samways, M.J. (1976) Song modification in the Orthoptera I. Proclamation songs of *Platycleis* spp. (Tettigoniidae), *Physiol. Entomol.* 1, 131–49.

Sanborne, P.M. (1982) Stridulation in *Merope tuber* (Mecoptera: Meropeidae), *Can. Ent.* 114, 177–80.

Saxena, K.N. and Kumar, H. (1984) Acoustic communication in the sexual behavior of the leafhopper, *Amrasca devastans, Physiol. Entomol.* 9, 77–86.

Schaffner, K.-H. and Koch, U.T. (1987) Effects of wing campaniform
    sensilla lesions on stridulation in crickets, *J. exp. Biol.* 129, 25–40.
Schilcher, F. von (1976a) The role of auditory stimuli in the courtship of
    Drosophila menalogaster, Anim. Behav. 24, 18–26.
Schilcher, F. von (1976b) The function of pulse song and sine song in the
    courtship of *Drosophila melanogaster, Anim. Behav.* 24, 622–5.
Schilcher, F. von (1976c) The behavior of *Cacophony*, a courtship song
    mutant in *Drosophila, Behav. Biol.* 17, 187–96.
Schilcher, F. von and Manning, A. (1975) Courtship song and mating speed
    in hybrids between *Drosophila melanogaster* and *Drosophila simulans,
    Behav. Genet.* 5, 395–404.
Schildberger, K. (1984) Multimodal interneurons in the cricket brain: Pro-
    perties of identified extrinsic mushroom body cells, *J. Comp. Physiol.*
    154, 71–9.
Schildberger, K. (1985) Recognition of temporal patterns by identified
    auditory neurons in the cricket brain. In K. Kalmring and N. Elsner
    (eds) *Acoustic and Vibrational Communication in Insects*, Berlin: Paul
    Parey.
Schuster, J.C. (1983) Acoustical signals of passalid beetles: complex reper-
    toires, *Fla. Entomol.* 66, 476–96.
Schwabe, J. (1906) Beitrage zur Morphologie und Histologie der tympanalen
    Sinnesapparate der Orthoptera, *Zoologica, Stuttgart* 20, 1–154.
Searcy, W.A. and Andersson, M. (1986) Sexual selection and the evolution
    of song, *Ann. Rev. Ecol. Syst.* 17, 507–33.
Selander, J. and Jansson, A. (1977) Sound production associated with mating
    behavior of the large pine weevil, *Hylobius abietis* (Coleoptera, Cur-
    culionidae), *Ann. Ent. Fenn.* 43, 66–75.
Selverston, A.I., Kleindienst, H. and Huber, F. (1985) Synaptic connectivity
    between cricket auditory interneurons as studied by selective photo-
    inactivation, *J. Neurosci.* 5, 1283–92.
Shaw, K.C. (1968) An analysis of the phonoresponses of males of the true
    katydid, *Pterophylla camellifolia* (Fabricius) (Orthoptera: Tettigoniidae),
    Behaviour 31, 204–60.
Shaw, K.C. (1974) Environmentally-induced modification of chirp length of
    males of the true katydid *Pterophylla camellifolia* (F.) (Orthoptera: Tet-
    tigoniidae), *Ann. Entomol. Soc. Am.* 68, 245–50.
Shaw, K.C., North, R.C. and Meixner, A.J. (1981) Movement and spacing
    of singing *Ambylcorypha parvipennis* males, *Ann. Entomol. Soc. Am.* 74,
    436–44.
Shorey, H.H. (1962) Nature of the sound produced by *Drosophila melanogas-
    ter* during courtship, *Science* 137, 677–8.
Shuvalov, V.F. and Popov, A.V. (1971) The reaction of females of the
    domestic cricket *Acheta domesticus* to sound signals and its changes in
    ontogenesis, *J. Evol. Biochim. Physiol.* 7, 612–6.
Shuvulov, V.F. and Popov, A.V. (1973) The importance of the calling song
    rhythmic pattern of males of the genus *Gryllus* for phonotaxis of
    females, *Zool. J.* 52, 1179–85.
Simmons, L.W. (1988) The calling song of the field cricket: *Gryllus bima-
    culatus* (De Geer): constraints of transmission and its role in intermale
    competition and female choice, *Anim. Behav.* 36, 372–9.
Simmons, P. and Young, D. (1978) The tymbal mechanism and song
    patterns of the bladder cicada, *Cystosoma saundersii, J. exp. Biol.* 76, 27–
    45.
Singh, B.N. and Chatterjee, S. (1987) Greater mating success of *Drosophila
    biarmipes* males possessing an apical dark black wing patch, *Ethology* 75,
    81–3.
Sismondo, E. (1980) Physical characteristics of drumming of *Meconema
    thalassinum, J. Insect Physiol.* 26, 209–12.
Sivinski, J. and Webb, J.C. (1986) Changes in a Caribbean fruit fly acoustic

signal with social situation (Diptera: Tephritidae), *Ann. Entomol. Soc. Am.* 79, 146–9.

Skaife, S.H. (1979) *African Insect Life.* London: Country Life.

Skovmand, O and Pederson, S.B. (1978) Tooth impact rate in the song of a shorthorned grasshopper: a parameter carrying specific behavioral information, *J. Comp. Physiol.* 124, 27–36.

Smit, F.G.A.M. (1981) The song of a flea – A stridulatory mechanism in Siphonaptera?, *Ent. Scand. Suppl.* 15, 171–2.

Smith, R.L. and Langley, W.M. (1978) Cicada stress sound: An assay of its effectiveness as a predator defense mechanism, *Southwest. Nat.* 23, 187–96.

Snodgrass, R.E. (1929) The thoracic mechanism of a Grasshopper and its antecedents, *Smith. Misc. Coll.* 82, 1–111.

Soper, R.S., Shewell, G.E. and Tyrrell, D. (1976) *Colcondamyia auditrox* Nov. sp. (Diptera: Scarcophagidae), a parasite which is attracted by the mating song of its host, *Okanagana rimosa* (Homoptera: Cicadidae), *Can. Entomol.* 108, 61–8.

Soucek, B. (1975) Model of alternating and aggressive communication with the example of kaydid chirping, *J. theor. Biol.* 52, 399–417.

Spangler, H.G. (1984) Silence as a defense against predatory bats in two species of calling insect, *Southwest. Nat.* 29, 481–8.

Spangler, H.G. (1986) Functional and temporal analysis of sound production in *Galleria mellonella* L. (Lepidoptera: Pyralidae), *J. Comp. Physiol.* 159, 751–6.

Spangler, H.G., Greenfield, M.D. and Takessian, A. (1984) Ultrasonic mate calling in the lesser wax moth, *Physiol. Entomol* 9, 87–95.

Spangler, H.G. and Manley, D.G. (1978) Sound associated with the mating behavior of a mutillid wasp, *Ann. Ent. Soc. Am.* 71, 389–92.

Speck, J. and Barth, F.G. (1982) Vibration sensitivity of pretarsal slit sensilla in the spider leg, *J. Comp. Physiol.* 148, 187–94.

Spieth, H.T. (1974) Courtship behavior in *Drosophila, Ann. Rev. Entomol.* 19, 385–405.

Spooner, J.D. (1968a) Pair-forming acoustic systems of phaneropterine katydids (Orthoptera, Tettigoniidae), *Anim. Behav.* 16, 197–212.

Spooner, J.D. (1968b) Collection of male phaneropterine katydids by imitating sounds of the female, *J. Georgia Entomol. Soc.* 3, 45–6.

Starck, J.M. von (1985) Stridulationsapparate einiger Spinnen-Morphologie und evolutionsbiologische Aspekte, *Z. Zool. Syst. Evolut.-forsch.* 23, 115–35.

Stephen, R.O. and Bailey, W.J. (1982) Bioacoustics of the ear of the bushcricket *Hemisaga* (Sagenae), *J. Acoust. Soc. Am.* 72, 13–25.

Stephen, R.O. and Bennet-Clark, H.C. (1982) The anatomical and mechanical basis of stimulation and frequency analysis in the locust ear, *J. exp. Biol.* 99, 279–314.

Stewart, K.W., Szczytko, S.W., Stark, B.P. and Zeigler, D.D. (1982) Drumming behavior of six North American Perlidae (Plecoptera) species, *Ann. Entomol. Soc. Am.* 75, 549–54.

Stout, J.F., DeHann, C.H. and McGhee, R.W. (1983) Attractiveness of the male *Acheta domesticus* calling song to females, *J. Comp. Physiol.* 153, 509–21.

Stout, J.F., Gerard, G. and Hasso, S. (1976) Sexual responsiveness mediated by the corpora allata and its relationship to phonotaxis in the female cricket, *Acheta domesticus* L., *J. Comp. Physiol.* 108, 1–9.

Suga, N. (1966) Ultrasonic production and its reception in some neotropical Tettigoniidae, *J. Insect Physiol.* 12, 1039–50.

Surlykke, A. and Gogala, M. (1986) Stridulation and hearing in the noctuid moth *Thecophora fovea* (Tr.), *J. Comp. Physiol.* 159, 267–73.

Szczytko, S.W. and Stewart, K.W. (1979) Drumming behavior of four nearctic *Isoperla* (Plecoptera) species, *Ann. Entomol. Soc. Am.* 72, 781–6.

Tautz, J. (1977) Reception of medium vibration by thoracic hairs of cater-
    pillars of *Barathra brassicae* L. (Lepidoptera, Noctuidae), *J. Comp.
    Physiol.* 118, 13–31.
Taylor, M.R. (1979) An acoustic stimulator for use with singing insects, *Lab.
    Pract.* 28, 137–8.
Theiss, J. (1982) Generation and radiation of sound by stridulating water
    insects as exemplified by the corixids, *Behav. Ecol. Sociobiol.* 10, 225–35.
Theiss, J. (1983) An acoustic duet is necessary for successful mating in
    Corixa dentipes, *Naturwissensch.* 70, 467–8.
Theiss, J. and Prager, J. (1984) Range of corixid sound signals in the
    biotope, *Physiol. Entomol.* 9, 107–14.
Thiele, D. and Bailey, W.J. (1980) The function of sound in male spacing
    behaviour in the bushcricket *Mygalopsis* (Orthoptera: Tettigoniidae),
    Austral. J. Ecol. 5, 275–86.
Thorpe, K.W. and Harrington, B.J. (1981) Sound production and courtship
    behavior in the seed bug *Ligyrocoris diffusus*, *Ann. Entomol. Soc. Am.* 74,
    369–73.
Thorson, J., Weber, T. and Huber, F. (1982) Auditory behavior of the
    cricket II. Simplicity of calling-song recognition in *Gryllus*, and anoma-
    lous phonotaxis at abnormal carrier frequencies, *J. Comp. Physiol.* 146,
    361–78.
Tischner, H. (1953) Über den Gehorsinn der Stechmucken, *Acoustica* 3,
    335–43.
Toms, R.B. (1985) Speciation in tree crickets (Gryllidae: Oecanthinae),
    Transvaal Mus. Monog. 4, 109–14.
Towne, W.F. (1985) Acoustic and visual cues in the dances of four honeybee
    species, *Behav. Ecol. Sociobiol.* 16, 185–7.
Trivers, R.L. (1972) Parental investment and sexual selection. In E.
    Campbell (ed.) *Sexual Selection and the Descent of Man: 1871–1971*, pp.
    136–79. Chicago: Aldine.
Turner, J.R.G. (1984) Mimicry: the palatability spectrum and its consequen-
    ces. In R.I. Vane-Wright and P.R. Ackery (eds) *The Biology of But-
    terflies*, pp. 141–62, London: Academic Press.
Tuttle, M.D. and Ryan, M.J. (1981) Bat predation and the evolution of frog
    vocalizations in the neotropics, *Science* 214, 677–8.
Uetz, G.W. and Stratton, G.E. (1982) Acoustic communication and re-
    productive isolation in spiders. In P.N. Witt and J.S. Rovner (eds)
    Spider Communication, pp. 123–59. Princeton, N.J.: University Press.
Ulagaraj, S.M. and Walker, T.J. (1973) Phonotaxis of crickets in flight: at-
    traction of male and female crickets to male calling songs, *Science* 182,
    1278–9.
Vrijer, P.W.F. de (1986) Species distinctiveness and variability of acoustic
    calling signals in the planthopper genus *Javesella* (Homoptera: Delphaci-
    dae), *Nether. J. Zool.* 36, 162–75.
Wadepuhl, M. (1983) Control of grasshopper singing behavior by the brain:
    responses to electrical stimulation, *Z. Tierpsychol.* 63, 173–200.
Waldron, I. (1964) Courtship sound production in two sympatric sibling
    Drosophila species, *Science* 144, 191–3.
Walker, T.J. (1957) Specificity in the response of female tree crickets (Or-
    thoptera, Gryllidae, Oecanthinae) to calling songs of the males, *Ann.
    Ent. Soc. Am.* 50, 626–36.
Walker, T.J. (1962) Cryptic species among sound-producing ensiferan Or-
    thoptera (Gryllidae and Tettigoniidae), *Quart. Rev. Biol.* 39, 345–55.
Walker, T.J. (1964a) Experimental demonstration of a cat locating orthop-
    teran prey by the prey's calling song, *Fla. Entomol.* 47, 163–5.
Walker, T.J. (1964b) Factors responsible for intraspecific variation in the
    calling songs of crickets, *Evolution* 16, 407–28.
Walker, T.J. (1969) Acoustic synchrony: two mechanisms in the snowy tree
    cricket, *Science* 166, 891–4.

Walker, T.J. (1979) Calling crickets (*Anurogryllus arboreus*) over pitfalls: females, males and predators, *Environ. Entomol.* 8, 441–3.

Wasserman, M. (1982) Evolution of the *repleta* group. In M. Ashburner and J.N. Thompson (eds) *The Genetics and Evolution of* Drosophila, Vol. 3b, pp. 61–139. London: Academic Press.

Webb, J.C., Burk, T. and Sivinski, J. (1983) Attraction of female Caribbean fruit flies *Anastrepha suspensa* (Diptera: Tephritidae) to the presence of males and male produced stimuli in field cages, *Ann. Entomol. Soc. Am.* 76, 996–8.

Weber, T., Thorson, J. and Huber, F. (1981) Auditory behaviour of the cricket. I. Dynamics of compensated walking and discrimination paradigms on the Kramer treadmill, *J. Comp. Physiol.* 141, 215–32.

Weidmann, S. and Keuper, A. (1987) Influence of vibratory signals on the phonotaxis of the gryllid *Gryllus bimaculatus* DeGeer (Ensifera: Gryllidae), *Oecologia* 74, 316–8.

West-Eberhard, M.J. (1984) Sexual selection, competitive communication and species-specific signals in insects. In T. Lewis (ed.) *Insect Communication*, pp. 283–324. London: Academic Press.

Weygoldt, R. (1977) Communication in crustaceans and arachnids. In T.A. Sebeok (ed.) *How Animals Communicate*, pp. 303–333, Indiana: University Press.

Wiese, K. (1981) Influence of vibration on cricket hearing: Interaction of low frequency vibration and acoustic stimuli in the omega neuron, *J. Comp. Physiol.* 143, 135–42.

Willis, L.J. (1946) British triassic scorpions. *Palaeontographical Soc.* Monograph, London.

Willmund, R. and Ewing, A. (1982) Visual signals in the courtship of *Drosophila melanogaster*, *Anim. Behav.* 30, 209–15.

Willows, A.O.D. (1968) Behavioral acts elicited by stimulation of single identifiable nerve cells. In F.D. Carlson (ed.) *Physiological and Biochemical Aspects of Nervous Integration*, pp. 217–44. Englewood Cliffs, New Jersey: Prentice-Hall.

Wilson, D.M. (1962) Bifunctional muscles in the thorax of grasshoppers, *J. exp. Biol.* 39, 669–77.

Wilson, D.M. (1964) The origin of the flight motor command in grasshoppers. In R.F. Reiss (ed.) *Neural Theory and Modelling*, pp. 331–45, Stanford: Stanford University Press.

Wilson, D.M. (1966) Central nervous mechanisms for the generation of rhythmic behaviour in arthropods, *Symp. Soc. Exp. Biol.* 20, 199–228.

Wohlers, D.W. and Huber, F. (1978) Intracellular recording and staining of cricket auditory interneurons (*Gryllus campestris* L., *Gryllus bimaculatus* DeGeer), *J. Comp. Physiol.* 127, 11–28.

Wohlers, D.W. and Huber, F. (1982) Processing of sound signals by six types of neurons in the prothoracic ganglion of the cricket, *Gryllus campestris* L. *J. Comp. Physiol.* 146, 161–73.

Wood, D. and Ringo, J.M. (1980) Male mating discrimination in *Drosophila melanogaster*, *D. simulans* and their hybrids, *Evolution* 34, 320–9.

Wyman, R.J. (1973) Neural circuits patterning dipteran flight motoneuron output. In J. Salanki (ed.) *Neurobiology of Invertebrates*. Budapest: Hungarian Academcy of Sciences.

Yager, D.D. and Hoy, R.R. (1986) The cyclopean ear: a new sense for the praying mantis, *Science* 231, 727–9.

Young, D. and Hill, K.G. (1977) Structure and function of the auditory system of the cicada, *Cytosoma saudersii*, *J. Comp. Physiol.* 117, 23–45.

Young, D. and Josephson, R.K. (1983a) Mechanisms of sound-production and muscle contraction kinetics in cicadas, *J. Comp. Physiol.* 152, 183–95.

Young, D. and Josephson, R.K. (1983b) Pure-tone songs in cicadas with special reference to the genus *Magicicada*, *J. Comp. Physiol.* 152, 197–207.

Zaretsky, M.D. (1972) Specificity of the calling song and short term changes in the phonotactic response by female crickets *Scapsipedius marginatus* (Gryllidae), *J. Comp. Physiol.* 124, 153–72.

Zeigler, D.D. and Stewart, K.W. (1977) Drumming behavior of eleven nearctic stonefly (Plecoptera) species, *Ann. Ent. Soc. Amer.* 70, 495–505.

Zeigler, D.D. and Stewart, K.W. (1986) Female response threshold of two stonefly (Plecoptera) species to computer-simulated and modified male drumming calls, *Anim. Behav.* 34, 929–31.

Zouros, E. (1981) The chromosomal basis for sexual isolation in two sibling species of *Drosophila: D. arizonensis* and *D. mojavensis*, *Genetics* 97, 703–18.

Zuk, M. (1987) Variability in attractiveness of male field crickets (Orthoptera: Gryllidae) to females, *Anim. Behav.* 35, 1240–8.

Systematic Index

Author Index

254

Subject Index